Junkyard Bandicoots
and Other Tales of the
World's Endangered Species

Other Titles of Interest from Wiley

Nature for the Very Young, Marcia Bowden
The Curiosity Club: Kids' Nature Activity Book, Allene Roberts
Get Growing, Lois Walker
National Gardening Association Guide to Kids' Gardening, Lynn Ocone with Eve Pranis
The Ocean Book, The Center for Marine Conservation
The Earth Book, Dinah Zike
Projects for a Healthy Planet, Shar Levine, Allison Grafton, and Terry Chui
Biology for Every Kid, Janice VanCleave
Chemistry for Every Kid, Janice VanCleave
Earth Science for Every Kid, Janice VanCleave
Physics for Every Kid, Janice VanCleave
Astronomy for Every Kid, Janice VanCleave
Math for Every Kid, Janice VanCleave
Looking at the Body, David Suzuki
Looking at Senses, David Suzuki
Looking at Weather, David Suzuki
Looking at Insects, David Suzuki
Looking at Plants, David Suzuki
Looking at the Environment, David Suzuki

Junkyard Bandicoots and Other Tales of the World's Endangered Species

Joyce Rogers Wolkomir and Richard Wolkomir

Illustrations by Laurie Davis

John Wiley & Sons, Inc.

NEW YORK · CHICHESTER · BRISBANE · TORONTO · SINGAPORE

Copyright © 1992 by Joyce Rogers Wolkomir and Richard Wolkomir

Published by John Wiley & Sons, Inc.

All rights reserved. Published simultaneously in Canada.

Photo reference for drawing on page 41 Duncan Willetts/Camerapix.

Library of Congress Cataloging-in-Publication Data

Wolkomir, Joyce.
    Junkyard bandicoots and other tales of the worlds endangered species / by Joyce and Richard Wolkomir.
      p. cm.
    Includes index.
    Summary: Discusses endangered animals around the world, including the razorback sucker and desert tortoise, and describes what can be done to save them.
    ISBN 0-471-57261-6 (paper)
    1. Endangered species—Juvenile literature. 2. Wildlife conservation—Juvenile literature. [1. Rare animals. 2. Wildlife conservation.]   I. Wolkomir, Richard. II. Title.
QL83.W65 1992
591.52'9—dc20   92-11114

Printed in the United States of America

10 9 8 7 6 5 4 3 2 1

Contents

The Tale of the Dodo: What It Means to Be Extinct

■ It is the year 1507. A Portuguese sailing ship has anchored off an unknown tropical island. The sailors lower boats and row ashore.

When they land on the beach, they find that nobody yet lives on this island, which is lush with bamboo thickets and coconut forests. At least, no *humans* live here.

Out of the jungle steps a huge, bluish-gray bird. It glares at the startled sailors—50 pounds of feathers, beak, and legs, standing as high as their chins.

No feathers grow on the bird's face. Its knobby beak is tipped in scarlet, and the bird seems to scowl. A tuft of feathers curls up on the bird's plump rear. It cannot fly—its wings are so tiny they look comical. Fluffy feathers on the bird's thick yellow legs make it seem to

be wearing pantaloons. To us, the bird might look funny, like Big Bird on "Sesame Street." But the enormous bird frightens the sailors.

They shoot it.

Later, they describe the bird's slowness and its lack of fear. They call it a "dodo," from a Portuguese word that means "simple-minded."

The island where the dodo lived is Mauritius, in the Indian Ocean. Eventually, people settled there. They cut down the dodos' forest home to plant sugar cane. Their dogs and hogs found the dodos' grass nests and ate the single large egg in each nest. Meanwhile, people killed dodos for food and for sport.

By 1681, the last dodo was dead.

Today, all that remains of the dodo are

DODO

kittens, and so they are a species. Daisy seeds grow into daisies, and so they are a species. We humans are a species. African lions are a species. So are monarch butterflies, white pine trees, hammerhead sharks, and gray squirrels. Scientists know of 1.4 million species. But they believe we know only a few of the Earth's animals and plants. They guess that the actual number of species now alive ranges from 5 to 30 million. Some species are big, some are tiny. They may be ugly or beautiful, edible, poisonous, quick, or slow. They may fly, crawl, swim, or spread their leaves in the sun.

But many of these species are now becoming extinct, like the dodo. They are disappearing at an alarming rate.

The dodo is one of 14 species that scientists know disappeared during the 1600s. In that one century, 14 species died off. But today, scientists believe species are becoming extinct at the rate of *one per hour*.

pieces. A head and foot are preserved at Oxford University, in Britain. A head is preserved in Copenhagen, Denmark. The British Museum has a foot. One American museum and six European museums have skeletons.

Dodos lived on the Earth for 180 million years. When dinosaurs roamed the world, so did dodos. But they could not cope with humans. Now the dodo is gone forever—extinct.

Vanishing Creatures

The dodos were a species. A *species* is any group of animals or plants that reproduces its own kind. Cats give birth to little cats, or

Where Are All the Creatures Going?

Animals and plants are disappearing because one species is different from the rest. That different species is us—the humans.

Usually, creatures keep each other in balance. For instance, mice have many offspring. They could become too numerous and eat all the plants and seeds. Then the mice would all starve. But hawks, snakes, weasels, and many other animals eat mice. That keeps the mice from becoming too numerous. Meanwhile, the mouse-eating animals generally have fewer young than mice do. They also prey on each other. Hawks eat snakes, for instance. In that way, creatures keep each other in check. Usually, no species becomes too numerous for its own good.

We humans are different. No other species regularly preys upon us. A shark or tiger occasionally attacks a person. Sometimes humans are attacked by other large *predators*—animals that hunt other animals. But that is unusual. Generally, no other species hunt humans as food. There is no natural check on humans, and so our population grows ever larger.

More people take up more room. And, unlike other creatures, we make enormous changes in the natural world.

Humans drain swamps and marshes to build houses. We burn rain forests to create cattle pastures. Our cities spread out, taking up more and more land. We also transform land by farming. For instance, the vast prairies of tall grass that once covered the center of North America have mostly been plowed into farms. As the prairies vanished, the great herds of bison that lived on the grasses have all but disappeared.

Each species must live in a community of certain types of animals and plants, called its *habitat*. This habitat may be on land or in the water. For instance, bison depended on the vast prairies with their tall grasses. Hunters

killed millions of bison. And when the prairies were gone, what remained of the great herds could no longer survive. Today, only a few bison remain, mostly in parks.

By destroying habitat, we doom many species. We also pour chemicals into the world. Our factories, power plants, cars, jets, and household furnaces emit fumes that contain chemicals. Some of these chemicals are poisonous to certain species. We also spray chemicals on our crops to keep insects from eating them. But some of these chemicals harm other creatures. For instance, certain insect poisons get into the eggs of some birds and weaken the shells. Because the shells easily crack, these birds cannot raise young. They begin to disappear.

Threatened and Endangered

As we destroy habitat and spew chemicals into the air, creatures are disappearing rapidly. When a species' numbers begin to fall, scientists say it is *threatened*. When a species has so few individuals left alive that it may not survive much longer, scientists call it *endangered*.

Sea otters are threatened. Rhinoceroses are endangered. Blue whales may be down to just 200 individuals. Fewer than 1,000 giant pandas are left alive. Unless our species does something, scientists say that within 10 years as many as half of all the species alive today may be gone.

In the chapters ahead, we will meet some

of these endangered and threatened animals. We will meet some of the Earth's largest creatures and some of its smallest. We will discover creatures that use their hair as weapons and creatures that "talk" among themselves in voices so low we cannot hear them at all.

We will learn how these creatures look and sound and where they live. We will learn why they are disappearing. And we will see what we humans can do—and are doing—to help them.

SECTION I

What Makes a Mammal a Mammal?

We are mammals. Angora cats and cocker spaniels are mammals. So are spider monkeys, grizzly bears, cottontail rabbits, pronghorn antelopes, and three-toed sloths. Blue whales and white-footed mice are mammals. We may be big or small. We may have two legs or four legs, or no legs at all, just fins. But we are all mammals. What do we have in common?

Our most important common trait is milk. Female mammals all have glands, called *mammary glands*, that produce milk to feed their young. Because young mammals receive nourishment directly from their mothers' bodies, mammals often have more family life than other classes of animal. Mammals tend to nurture their young longer and to teach their young more about how to behave and make their way in the world.

Mammals also have hair. Scientists believe that mammals began millions of years ago as small dinosaurs. Between their scales, these reptiles had tiny, hairlike projections that enabled them to feel their surroundings, even though their scales acted as coats of armor. As the reptiles slowly evolved into mammals, they lost their scales, although remnants of them are still visible here and there, such as on the scaly tails of rats. Meanwhile, the projections growing between the scales evolved into hairs. At first, these hairs were for feeling things, as animal whisker hairs still are. Eventually, though, the mammals evolved other uses for their hair. For instance, they use their hair as weapons: porcupine quills are evolved hairs and rhinoceros horns are compressed hair. Also, hair became coats of fur, which keep animals warm by trapping a layer of heated air next to their skin.

Mammals are all warm-blooded, which means we can more or less keep our inner temperature the same, even if the outside air is much colder. Birds also are warm-blooded. But many other classes of animal, such as reptiles, are cold-blooded: if the air gets cold, so do they.

We mammals number about 4,000 species, a varied group. Australian platypuses lay eggs. Kangaroos and opossums raise their young in pouches. The blue whale is the largest animal that ever lived, and the pygmy shrew is tiny. Bats fly like birds; dolphins swim like fish. We humans have lost most of our hair, walk on two legs, and operate computers. But we all are mammals.

Will We Hear the Elephants?

A woman is visiting the elephants at the zoo in Portland, Oregon. Seemingly, the elephants in the enclosure are silent. But she wonders, Could the elephants be "talking" in voices she cannot hear?

It is an afternoon in 1984. The woman is Katy Payne, a Cornell University expert on animal sounds. She feels something strange in the elephant enclosure—a throbbing in the air.

"It was like the vibration you feel when a very large organ hits the lowest notes," she remembers today.

She wondered if the elephants could be rumbling to each other in voices too low for humans to hear. To find out, she brought microphones and recording gear to the zoo.

Working with another Cornell researcher, Bill Langbauer, she discovered that elephants do "speak" to each other in low rumbles.

Elephants make many sounds that humans *can* hear: barks, snorts, trumpets, roars, and growls. To find out why elephants also make their ultra-low rumbles, Payne and Langbauer joined forces with African elephant specialist Joyce Poole.

Riding a jeep among the elephant herds in southwestern Africa, the researchers used electronic gear to record the elephants' too-low-to-hear rumbles. On other expeditions, they set up their recording devices in towers. Over time, they learned what rumbles elephants make and why they make them.

For one thing, these low-pitched sounds travel long distances. Even dense forests do not stop them. The researchers discovered that elephants hear the rumbles of friends several miles away. And so the calls are an elephant long-distance telephone system, allowing the animals to communicate with each other even when they are far apart and out of sight.

Groups of elephants coordinate their

ASIAN ELEPHANT

AFRICAN ELEPHANT

movements with the rumbles, even when they are separated. Also, elephants use their low calls to attract mates. Elephants communicate often with each other because they have a complex social life.

Female elephants live in groups of their relatives. Females stay in their group, but males leave after they are about 15 years old. Then the males spend part of their time in groups of other males and part of their time wandering on their own. Elephants live about 65 years.

Elephants know each other as individuals. Mothers teach their daughters and sons where to find water and food. Cynthia Moss, director of the Amboseli Elephant Research Project in Kenya, Africa, has found that elephants even seem—in some way—to understand death.

"Unlike other animals, elephants recognize one of their own carcasses or skeletons," she has written. Moss says that elephant mothers also show signs of being depressed if their calf dies, looking lethargic and trailing far behind the herd.

Thousands of years ago, during the Ice Age, huge elephants called woolly mammoths and mastodons roamed Europe and the Americas. But they became extinct. Scientists do not yet know why. Today, only two elephant species remain: Asian and African. African elephants are slightly larger, with much bigger ears. Otherwise, the two species look and act much the same, although they live on different continents. Both species are in danger of extinction.

One reason elephants are in trouble is that the growing human population of Africa and Asia is eliminating their habitat. They need a lot of land to roam because each elephant eats up to 500 pounds of leaves and twigs a day and drinks up to 40 gallons of water at a time. An elephant uses its trunk—which is a combination of its upper lip and nose—as a sensitive "hand." An elephant also can use its trunk as a hose: it draws water into its trunk, then curls the trunk down to spray the water into its mouth.

Elephants also are endangered because humans use the ivory from their huge tusks to carve figurines. It is illegal to hunt elephants. But *poachers*—people who kill or capture animals illegally—sneak into game preserves. They shoot the elephants and cut off their tusks to sell. As a result, the number of elephants has been dwindling. Africa had over a million elephants 10 years ago; today, only about half that number remain. In Asia, only 40,000 to 60,000 elephants are still alive, and most of them are captives, used as beasts of burden to carry and pull heavy loads.

Female Asian elephants have no tusks, unlike African elephant females. And the tusks of male Asian elephants are smaller and less valuable than African elephant tusks. As a result, elephant poaching is rarer in Asia.

In Africa, there is now hope: many nations around the world have banned the import of ivory, with dramatic results. In the late 1980s, before the ban went into effect, poachers in Kenya alone killed 5,000 elephants a year. In 1990, after the ban, poachers killed only 55 elephants in Kenya's game parks.

Nevertheless, the number of elephants is dwindling. Conservationists must find new ways for people and elephants to live together. Then elephants may continue to wander, calling to each other in voices we cannot hear.

Junkyard Bandicoots Find Refuge

Eastern-barred bandicoots are making their last stand in an Australian junkyard. Scooting among thrown-away soda cans and auto engines, the little mammals look like large rats with black racing stripes on their rumps.

Actually, bandicoots—like kangaroos—are a kind of mammal called *marsupials*, which carry their young in a pouch. Bandicoots come out at night. The eastern-barred bandicoot tunnels into the ground with its long, pointed

EASTERN-BARRED BANDICOOT

snout, searching for insects and plants. Once Australia teemed with bandicoots; the continent has 22 species. But farmers tried to exterminate them as pests. Development, plus competition for food from sheep and rabbits, also has pushed them close to extinction.

Australia's few hundred remaining eastern-barred bandicoots found refuge under old car bodies in a junkyard in Hamilton, in Australia's southeast. Now they are benefiting from human kindness. Hamilton residents built bandicoot homes in the junkyard from railroad ties, concrete culverts, and sheets of tin.

Hamiltonians also tied bells on their cats' collars. Now bandicoots can hear the cats coming and scurry away.

Big Guy Is Helping His Species

Big Guy lives inside a pen in Florida. He is watched over by Dr. Melody Roelke, a veterinarian with the Florida Game and Fresh Water Fish Commission. He receives rich meat, but he can no longer live in the wild.

Big Guy is a Florida panther. Sometimes Big Guy meows like a house cat, only much louder. A full-grown Florida panther measures up to seven feet from its nose to its tail. Its kittens are born with bright blue eyes. As they grow, their eyes turn golden brown.

Once, these panthers were the U.S. South's top predators—animals that kill and eat other animals. They are hunters of large prey, particularly deer and wild hogs. But European settlers chopped down the South's forests to create pastures. They drained swamps to grow cotton. As the habitat vanished, so did the deer and wild hogs that the panthers ate.

Sometimes hungry panthers attacked settlers' livestock. Settlers even spread tales of ferocious panther attacks on people. The state offered money for every panther killed, and hunters shot every panther they saw. By the late 1920s, the few panthers left alive had retreated to the wild swamps of southern Florida. With the forests gone, the panthers had nowhere else to go. But even the swamps have not been a safe haven.

Although panthers are protected now, fewer than 50 are still alive. And they have a new enemy: the automobile.

Highways cut through the Everglades and the Big Cypress Swamp in southern Florida. Every day, thousands of cars and trucks rush along these highways. In the past 15 years, cars have hit at least 20 panthers. Of these, four are known to have survived. One was Big Guy.

At 6:00 A.M. one day, a truck driver found the big cat beside the road. He was badly hurt after being hit by a car. The trucker went for help. Wildlife officials airlifted Big

FLORIDA PANTHER

Guy to the University of Florida, where veterinarians saved his life. They repaired broken bones in his rear legs with metal plates. While his legs were healing, Big Guy broke off three teeth when he tried to bite a metal bar. His caretakers repaired his teeth, but feared the teeth would not last in the wild. And so he now lives in a northern Florida preserve.

Big Guy can never return to the wild himself. Even so, he is helping wildlife officials who are trying to save his species. He is part of a program to breed panthers in captivity for eventual release into the wild.

Such programs are necessary because wild panthers face a grim future. Cars are a danger.

Also, with all-terrain vehicles and swamp buggies, people can follow panthers into the thickets. That makes it difficult for them to hide and to hunt. Meanwhile, the panthers' last bit of habitat is getting smaller.

For nearly a century people have been slowly draining the southern Florida swamps, turning the wetlands into farmlands and housing developments. People also have hunted deer and wild hogs for sport, leaving panthers in some areas little large game to eat.

"These are large predators and they need large animals for food—deer and hogs," says Dr. Melody Roelke. "Now—in some areas—they're eating raccoons and armadillos, and they're not getting enough to eat."

Wildlife biologists keep a close watch on the panthers. They fit panthers with collars that send out a radio signal. Then they fly over the swamps in airplanes. By tuning into the signals from the radio collars, the scientists in the airplanes can follow the panthers as they prowl through the swamp.

The scientists also hike into the swamps to find panther kittens. They give the kittens vaccines to protect them against diseases and vitamins to keep them healthy.

Meanwhile, Florida has fenced off a section of interstate highway where the panthers used to cross and built underpasses so that panthers can cross the highway safely. The state has limited the use of all-terrain vehicles, although swamp buggies still roar through the wetlands. Florida also has restricted hunting of deer and wild hogs in some—but not *all*—areas where panthers live, so they will have prey.

Now scientists are trying to help the panthers more directly. They have begun to capture wild panther kittens to raise in pro-tected sanctuaries. They hope to breed the captive panthers. They are also using panthers like Big Guy, who can no longer live in the wild. Eventually, the scientists will send some of the panthers they raise to zoos, so that even if the panthers die out in the wild the species will still exist in captivity. But their main goal is to release panthers raised in captivity back into the wild. That way, Big Guy's children may prowl his former home in southern Florida's swamps.

The problem is that so little of the wild is left. Florida has established one panther wildlife refuge. Dr. Melody Roelke and her coworkers on the captive breeding project hope to find wild spots in other southern states where they can release the panthers they raise in captivity. That way, the species would not have to depend on one site for survival.

Scientists are doing what they can. But they can only hope that Big Guy will not be one of the last of his kind.

4

Return of the Black-Footed Ferret

In 1981, a dog that lived on a Wyoming ranch trotted home carrying a dead black-footed ferret. But the ferrets—cousins of the weasel and mink—were supposed to be extinct!

Wildlife biologists searched the prairie near the ranch, looking for more ferrets. They are dachshund-sized, long and sinewy, with a gray body and black legs. Their faces are white with a black bandit's mask. And the biologists knew where to look for them: black-footed ferrets live in underground "towns" dug by prairie dogs.

Prairie dogs are small relatives of squirrels. They burrow under the prairie grasses. Hundreds dig their burrows together, forming a "town." Before European settlers came, prairie-dog towns covered many acres, with entrances and lookout mounds poking up through the tall grass. Prairie dogs shared their burrows with other animals. The town was an *ecological system*, a place where many plants and animals live together, relying on each other.

Black-footed ferrets ran through the burrows, hunting prairie dogs to eat. The ferrets had big eyes, which allowed them to see in the dark. But they also had sensitive whiskers on their muzzles and on their legs to feel their way through the burrows. Ferrets helped prevent prairie-dog population explosions. So did other predators, like rattlesnakes, bull snakes, kit foxes, golden eagles, and prairie falcons. Meanwhile, the ferrets themselves were prey for larger hunters, like badgers, great-horned owls, and coyotes.

Farmers and ranchers believed that the prairie dogs ate grasses their livestock needed. Today, many farmers and ranchers still believe prairie dogs are their enemies. Conservationists disagree. They say studies show that cattle grazing on a prairie-dog town weigh just as much as cattle grazing where there are no prairie dogs. But settlers declared war on prairie dogs. Around 1900, the U.S. government began to poison prairie dogs out of existence.

Today, about 95 percent of the prairie-dog towns are gone, and the poisoning continues. As the prairie dogs dwindle, so do many creatures that depended on them.

One victim was the black-footed ferret. By 1981, biologists believed none were left. But that was the year the Wyoming ranch dog trotted home with a dead ferret in its mouth.

Near the ranch, biologists found a prairie-dog town that was home to some 60 ferrets. The black-footed ferret was not extinct after all. However, it was one of the Earth's most endangered mammals. Then, in 1985, a disease called distemper struck the ferrets. A year later, only 18 were left alive.

Biologists held an emergency meeting. They could leave the ferrets in their burrows. But then the disease would probably kill them all. Reluctantly, the biologists decided to take the ferrets into captivity, where they would be safer.

It proved wise. The captive ferrets lived in a center designed by the Wyoming Game and Fish Department. Biologists limited human contact with the animals so they would remain wild. As the ferrets reproduced, the biologists sent some to a Smithsonian Institution research facility in Front Royal, Virginia, and to several zoos. That way, if an epidemic broke out among one group of captive ferrets, others would be safe. By 1991, the ferret population had grown to 250 and biologists believed it was time to begin returning the species to the wild.

To do that, the scientists had to negotiate with ranchers. The ranchers feared that saving the ferrets would mean restrictions on their use of private and public land for grazing livestock. Others worried they would no

BLACK-FOOTED FERRET

longer be able to drive all-terrain vehicles over prairie-dog towns. Finally, the biologists persuaded people that the ferrets would not force them to change their activities.

On September 3, 1991, the biologists released two ferrets near the ranch where they were rediscovered in 1981. As the two ferrets ran out onto the prairie, Wyoming governor Mike Sullivan quoted from the Bible: "Go forth and multiply."

By November, 1991, the biologists had released 49 ferrets. A badger and coyotes killed five of the released ferrets. But the biologists expected that up to 90 percent of the released ferrets would die of natural causes. They plan to keep on breeding captive ferrets and releasing them in the wild, first in Wyoming and eventually in other western states.

Hopes are high. With luck and with help from humans, black-footed ferrets may return to the American prairie.

HABITAT CLOSE-UP

PRAIRIE

When European settlers first made their way into what are now the midwestern states, they discovered that the eastern forests dwindled away. Ahead, they saw only a vast sea of grass: the American prairie.

Prairie is level grassland. At first, it seemed empty, just wind-blown grass with the sky above. But the prairie was alive.

The rich soils produced grasses taller than a basketball player. Between the green grasses grew flowers, and these plants were food for large bison.

Perhaps 60 million bison roamed the prairies. These herds could pound down the soil and make it too dense for the very grasses they ate to grow. But tiny burrowing animals—ground squirrels, gophers, mice, and voles—loosened the soil, letting in air so the grasses could grow. They also ate grass. Too many of them might eat *all* the grass. But predators like hawks, snakes, coyotes, badgers, and ferrets kept their numbers down.

Not only prairie dogs lived in the vast prairie-dog "towns." In the winter, box turtles, lizards, and toads all sheltered in the burrows. In the spring, burrowing owls nested there. And in the summer, all sorts of birds lived around the prairie-dog towns, from lark buntings to mountain plovers.

Now the prairies have mostly been plowed into farms growing corn and wheat. But the grasses and other prairie plants still grow in patches along railroad lines and the sides of highways.

Today, some people are trying to set aside stretches of land for the prairie plants. Someday, we may see the prairie grasses waving in the wind again, as they did a hundred years ago.

5

For Pandas, Being Loved Is Not Enough

People love pandas. They look like white bears with black legs and black eye patches. People line up at zoos to see them.

But zoos may soon be the *only* places to see pandas. In the wild, they are disappearing fast.

Pandas live only in the mountains of southwestern China. They are disappearing because their habitat is disappearing. Also. people shoot them for their fur.

A Chinese hunter can sell a panda pelt for as much as $20,000 in Taiwan and Japan. China has made the punishment for shooting pandas death. Even so, hunters continue to shoot pandas illegally—poaching. So far, at

least four poachers have been sentenced to death. But the shooting goes on.

About 100 pandas now live in zoos around the world. In the wild in China, fewer than 1,000 pandas are left alive. One reason scientists have difficulty helping pandas is that these animals can live in just one habitat.

Pandas eat only bamboo that grows in certain Chinese mountains. To get enough food from bamboo, a panda must eat all day. But China has many people and they cut down ever more of the bamboo forests for fuel and to create farmlands. In the past 20 years, pandas have lost 40 percent of their habitat. China has set up a handful of panda reserves, but people continue to cut bamboo even in the reserves. At one reserve, the Chinese authorities have built a power station hoping it will keep the pandas' neighbors from chopping down their bamboo for fuel.

Wildlife officials try to breed pandas in captivity and return them to the wild. But the plan has not worked well. For one thing, pandas have few babies. Also, when they do have babies in zoos, the little pandas are so tiny and the mother pandas are so huge that sometimes the mother accidently hurts her

GIANT PANDA

baby. In 20 years of trying to breed pandas in captivity, Chinese panda experts have raised only 28 pandas. Making the problem of protecting pandas worse, say Chinese panda experts, is that the several government agencies responsible for protecting pandas do not cooperate.

Right now, the worst threat to pandas is poaching. Unless something changes, say biologists, pandas could be extinct in the wild within 20 years.

As one Chinese expert recently told an American visitor: "Our hearts are aching with anxiety—we know the panda must rely on man to survive. But man has not yet offered a good way of helping it."

6

Lemurs Are Leaping for Their Lives

In a bamboo forest on the island of Madagascar an animal chattered. Patricia Chapple Wright held her breath, listening.

Wright was an anthropologist at the Duke University Primate Center in North Carolina. She had come to Madagascar to find a creature most scientists believed extinct: the greater bamboo lemur. A relative of the monkeys, the greater bamboo lemur had not been seen for 15 years, since 1972. But Wright hoped some of the lemurs might still be hiding in Madagascar's forests of giant bamboo, the lemurs' food.

From a village of tin-roofed huts, she and her team of researchers climbed to the mountaintops. Below, lumberers had cut down all the trees. But on these high slopes, patches of ancient forests remained. Orchids bloomed in trees covered with red moss. And beside a stream foaming down rocky cliffs grew giant bamboo. The bamboo grew 65 feet high, with stalks as thick as a man's wrist. After weeks of searching in the rain, Wright heard an animal scold. She froze, neither breathing nor blinking.

A lemur was scowling at her, clutching a bamboo stalk. It seemed part squirrel, part monkey. It was furry red, with chubby, golden cheeks. It had teddy-bear ears and a long orange tail.

"Chuck-chuck," the lemur scolded. Then it twirled its long tail and leaped out of sight.

Wright rushed through the bamboo, shouting to the other researchers: "It's velvety red! It's beautiful! It's not extinct!"

A week later, four stories above her in the foliage, she spotted two more greater bamboo lemurs, sunning themselves. They glided away into the trees. An hour later she found them again. But now there were four, sitting in a row on a branch: two adults, an infant, and a "teenager" chewing bamboo stems.

Wright called the four lemurs her "gang."

She found a tall palm tree atop a ridge where the family climbed at dusk to sleep, each using a separate palm frond as a backrest.

After a month of studying the "gang," Wright decided to look elsewhere on the island for more greater bamboo lemurs. She and her associates drove through miles of grassy hills that had once been rain forest, cut down for lumber. At each village she asked the local leader, or "president," about giant bamboo and lemurs. The answer was always the same: "*Tsy misy. Tsy misy*," which means "There are none. They are no longer here."

Millions of years ago, Madagascar was a chunk of Africa that broke off and drifted away. The island's creatures evolved on their own. As a result, Madagascar has many species that live nowhere else, like the lemurs.

Lemurs are relatives of monkeys and apes. They are our own distant relatives, too. Some are as small as chipmunks and some are as big as cocker spaniels, with long, bushy tails. All are unusual. One lemur, the aye-aye, has large round ears and big round eyes. On each hand, its middle finger is unusually long. The aye-aye uses its long finger to tap tree trunks: somehow the tapping tells the aye-aye if there is a cavity under the wood.

Scientists are not sure if the aye-aye detects cavities by the sound of its tapping or

LORIS

RING-TAILED LEMUR

by the way the wood feels. But the aye-aye also can tell if the cavity is empty or if it contains an insect the aye-aye can eat. It is the only mammal known to find its food by tapping like a woodpecker.

The aye-aye, like so many other Madagascar species, is endangered. That is because the island's population has more than doubled over the last 40 years. The average income is only a few hundred dollars per year. To support themselves, people are cutting down their rain forests to sell the wood as lumber or burn it for fuel. Vast sections of the land are now bare. As the barren hills erode, streams run red with mud. Astronauts looking down from space have seen a red ring of washed-away soil around the island.

Scientists from around the world are trying to help Madagascar save what is left of its unique plants and animals. That is good news for the greater bamboo lemur: in the spring of 1991, the government of Madagascar protected the mountaintop forest where Patricia Wright discovered her "gang." It is Madagascar's newest national park.

For the 80,000 people who live around the park and need its wood, however, the preserve could make life difficult. But the American forester who heads the new park is trying to turn it into a resource for nearby residents. For one thing, scientists flock to the new preserve to study its rare creatures, providing neighboring villagers with new jobs as guides. Increasing numbers of tourists should create a need for more guides.

Eventually, further ways must be found for Madagascar's people to survive without cutting down the remaining forests. For now, the greater bamboo lemur—back from extinction—has a chance for a future.

7

Manatees Are on a Collision Course with Boats

At Homosassa Springs State Wildlife Park in Florida, a ranger in a wet suit wades into a spring-fed pool. Several adult manatees and a baby manatee gather convivially around her, ready for lunch.

They are charming animals: tubby blobs with flippers instead of legs and good-natured faces. They grow as long as rowboats and weigh up to two tons. Although they resemble seals in need of a diet, they are relatives of elephants. But their *spiritual* cousins are those gentle creatures of the meadows—cows.

"Sea cow" is the manatee's nickname, for it spends its days quietly grazing on underwater plants, harming no one. Manatees may have inspired the ancient legends about mermaids. Despite all their blubber, they are beautiful because of the gentleness of their eyes and their good nature.

As the Homosassa Springs ranger holds up lettuce and carrots, the manatees rub her with their muzzles, a greeting. Florida swimmers sometimes are astonished when a huge manatee paddles up to give them a gentle hug and a manatee kiss. The Homosassa manatees, each in its own way, request a carrot: one nuzzles the ranger's wet suit, another wiggles its upper lip, and a third—to make sure there is no doubt—rolls on his back and points his flipper at his open mouth.

Within a decade or so, manatees are likely to be extinct. They have been legally protected for nearly a century. Yet, only about 1,000 of the Florida manatees remain.

In part, the species is fading because of habitat loss. The manatee's hugeness helps to keep it warm. But to maintain its enormous weight, each manatee must eat 100 pounds of sea grasses, water hyacinths, and other plants every day. As pollution and development thin the plants, manatees have less to eat. However, their biggest enemy is boats.

Florida may be the world's boating capital. In 20 years, the state's boat population

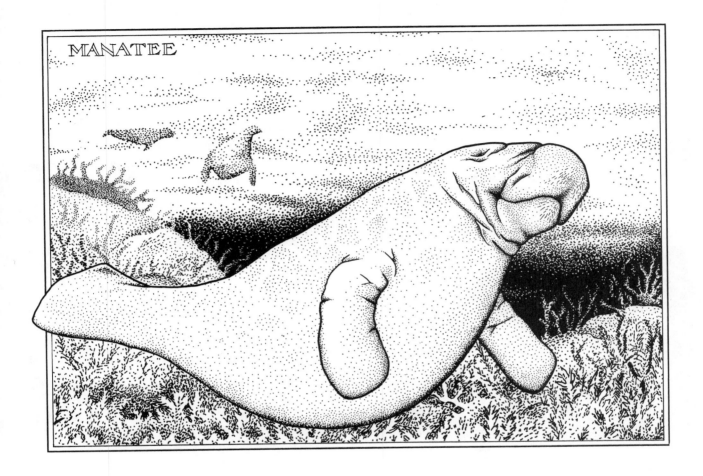

MANATEE

has more than tripled: over 700,000 boats are registered in the state, with up to 300,000 "visitor" boats annually. Boaters like to whoosh at 40 miles per hour through bays and inlets where manatees graze on bottom grasses, occasionally drifting to the surface to unflap their nostrils and breathe. Too often, a surfacing manatee is hit by a speedboat; about 85 percent have scars from propeller gashes and boat collisions.

The U.S. Fish and Wildlife Service studied manatee deaths over 15 years. Of the 510 deaths from known causes, 418 were from boat or barge collisions. Others died in canal locks. Some were entangled in fish nets. And some were senselessly shot by vandals despite the $20,000 fine. In January, 1990, abnormal cold killed 74 manatees. But it is chiefly boat collisions that reduce the manatee population by about 20 percent each year.

Recovering from such losses is difficult. While individual manatees may live to be 60 years old, they produce only a single pup every two or three years. And infant mortality is rising, apparently because water pollution taints their mothers' milk.

Florida manatee advocates are trying desperately to save the species. For instance, they have pushed new state restrictions that require boats to slow down in waters where manatees live. Recently, concern over the safety of manatees led federal officials to cancel a championship professional powerboat race at Hillsborough Bay. Florida is planning to step up enforcement of laws protecting the species and to control the expansion of boating facilities in manatee areas. Meanwhile, at a few refuges like Homosassa Springs Wildlife Park, the mammals can live and breed in safety.

8

When Gray Wolves Howl

In a Minnesota forest, wolves are howling. Moonlight glitters on snow-covered spruces and pines as the wolves raise their voices. What the wolves are "saying" depends on who is listening.

A farmer hears, "We will eat your sheep." A mother in a house at the forest's edge thinks the wolves say, "We may attack your child." A couple skiing through the forest stop and smile. To them, the wolves' song is beautiful: "We are wildness."

A biologist is listening, too. He hears the wolves say, "Keep away you other wolves— this is *our* area!"

Wolves say many things when they howl, scientists have found. Sometimes they call their friends: "I'm over here." Sometimes they just say, "Isn't it fun making all this noise?"

Scientists are studying wolf howling as one way to learn how wolves live. Wolves are endangered in the United States (except Alaska) and in other parts of the world, too. By learning about wolves, scientists hope to help them. Also, as we've found out more about wolves, more people now admire them and want to know more about them. It is a big change.

Wolves were in the world before humans. Some scientists believe our distant ancestors learned to hunt by watching wolves, which are clever hunters. In fact, early human societies may have resembled wolf packs. Scientists know that our dogs—from dachshunds to Saint Bernards—are all descended from wolves. But for most of history, humans have hated and feared wolves. As soon as Europeans arrived in North America they began killing wolves, which used to live everywhere on the continent. Today, south of the Canadian border, virtually the only wild wolves left are in northern Minnesota. Yet, many people still hate wolves.

Farmers and ranchers fear wolves will kill their valuable livestock. And old legends tell of wolves attacking people—some still believe wolves are dangerous. But biologists studying wolves find they are not really "bad guys."

GRAY WOLF

Wolves live in "packs," which are complex communities not unlike our own. Usually, a pack is a family, consisting of a father, a mother, some "teenagers," some pups, and a few aunts and uncles. Every wolf has a place in the pack, just as every soldier has a place in an army.

The pack's commander is the biggest, oldest, wisest male, called the "alpha" wolf. His mate is the top female. Second in command is the "beta" wolf, who may lead hunting expeditions. All the wolves know their "rank," whether they are the equivalent of sergeants, corporals, or privates. Wolves have ways—such as how they carry their tails—of showing their rank in the pack. Every wolf knows his or her place in the pack and follows the "orders" of the alpha or beta wolf as they run after prey. This makes the wolf pack highly efficient at hunting.

Only the alpha wolf and his mate have young. The other wolves in the pack help bring food to the pups, who spend their first few weeks hidden in a den dug into the ground. Sometimes wolves will "babysit" with the pups so their mother can run with the pack for a while. Pups spend their days eating and wrestling with each other, growing strong so they can take their place in the hunt.

Scientists have carefully studied the relationship between wolves and the animals upon which they prey, such as deer or moose. They have found that wolves usually run down and kill prey animals that are old, sick, weak, or not yet grown. That may seem cruel, but it allows the strongest of the prey animals to survive and pass on their traits to their babies. That way, the herd grows stronger. Human hunters, on the other hand, tend to shoot the biggest and "best" deer or elk, which may weaken the herd.

Also, without wolves to keep their numbers in check, deer, moose, or elk may have a population explosion. They become so numerous that they eat all the available food. Then a major portion of the herd dies of starvation. Generally, predators and prey keep each other in balance.

In studies in Minnesota and elsewhere, biologists have found that wolves *do* sometimes attack farm animals, especially when their normal prey is gone. But such attacks are rare. In Minnesota, wolves kill one cow in every 2,000 cows and one sheep per 1,000 sheep. Funds set up by conservationists help repay the few farmers or ranchers who lose livestock to wolves.

Studies also show that humans have almost nothing to fear from wolves. In the history of the United States, there has never been a recorded instance of a wild wolf killing a human. All animals must be avoided if they have the disease rabies, which affects the brain. This includes wolves, but rabid wolves are rare.

Slowly, many people are coming to appreciate the wild beauty of wolves. As a result, there is now a chance for wolves to return to at least a few places in the United States where they are currently extinct or endangered.

Alaska has about 6,500 wolves. Perhaps 50,000 wolves live in Canada. In the continental United States, however, the only real population is in Minnesota, which has about 1,600 wild gray wolves. A few packs live on Michigan's Isle Royale. A small number of Minnesota wolves recently moved into Wisconsin. A few Canadian wolves have pushed southward into Montana. Meanwhile, conservationists are hoping to help wolves settle in other parts of the United States.

One plan calls for returning gray wolves to Yellowstone National Park. If wolves are reintroduced there, Yellowstone will have the same animals it had when Columbus first landed in America.

North America has two kinds of wolves. One is the gray wolf, which lives in the north. The other, the southern red wolf, has been extinct in the wild since 1975. Now biologists hope captive red wolves released in North Carolina will survive in the Great Smoky Mountains National Park. They also hope to return a southwestern type of gray wolf, the "Mexican wolf," to New Mexico.

Nobody knows if the plans will succeed. If they do, Americans once again will hear the howling of wild wolves.

9

Monkeys That Go to Survival School

A monkey the size of a squirrel steps warily from a box. It has just arrived in Brazil from a zoo in the United States. Its black eyes are round: it sees trees and more trees. Never has the monkey seen so many trees.

The yellowish-orange monkey has a hairy mane. It looks like a tiny lion. It is called a "golden lion tamarin." Its ancestors once trooped through the lush foliage of this rain forest along the Brazilian coast. But now the forest is almost gone, chopped down for farms. Just patches are left.

As the forest has vanished, so have the golden lion tamarins. Only about 400 are left in the wild.

In the early 1980s, the U.S. National Zoo in Washington, D.C., began a save-the-tamarins program. A hundred zoos around the world joined the effort to breed golden lion

tamarins in captivity. The aim is to return the zoo-bred monkeys to tropical forest preserves in Brazil. But zoo-raised golden lion tamarins do not know how to *act* like golden lion tamarins; they have never seen a rain forest.

Wild golden lion tamarins learn from their parents to find food and avoid danger. But for zoo-raised monkeys, like the frightened tamarin just arrived from the United States, the jungle is baffling.

That is why the newcomer must attend "survival school." The "school" is in a cage at the forest's edge.

The "students" learn to find insects and fruit the biologists have hidden in rolled-up tree leaves or in bark crevices. They practice swinging on vines and branches, learning how large a branch must be to bear their weight. They must even learn how to peel bananas.

"Graduates" of the survival school return as "teachers." By watching these rain-forest veterans, the newcomers learn to find their way through the tangled treetops. They learn to watch out for snakes and to tell good fruits from poisonous fruits. Eventually, the zoo tamarins leave their cage and enter the real rain forest. Biologists still watch over them, but they are on their own. Many do not survive the first year, for they still have much to learn.

Released monkeys must learn to live as families, because both father and mother tamarins help to raise the young. Older brothers and sisters "babysit." A tamarin family sleeps in a huddle, in a hole in a tree trunk or in a nest of intertwined branches and vines. When a hawk glides overhead, a tamarin flattens itself against a tree trunk, trying to disappear. If a predator threatens from the ground—a snake, perhaps, or an ocelot— tamarins join forces to attack the predator and drive it away.

One enemy of the tamarins is humans. People sneak into the sanctuaries and steal the released tamarins to sell as pets. Golden lion tamarins are so rare that a single monkey can sell for as much as $20,000.

Meanwhile, all that remains of Brazil's coastal rain forest is patches atop hills and mountains. Conservationists are working to maintain rain-forest preserves, where creatures like the golden lion tamarin can live.

For the zoo-bred newcomer, just off the airplane in the Brazilian rain forest, the odds are against lasting through its first year in the wild. But if the tamarin pays attention in survival school, it has at least a chance of becoming a wild monkey, living free in the treetops. If enough make it back into the wild and reproduce, the golden lion tamarin may have a future.

GOLDEN LION TAMARIN

HABITAT CLOSE-UP

No habitat on Earth is richer in life than a tropical rain forest. It rains here virtually every day, and it is always warm. All sorts of plants flourish year round, providing food for thousands of different insects and animals. And those plant eaters, in turn, become meals for predators like eagles, hawks, cats, and snakes that eat smaller creatures. Life in the rain forest is rich and complicated.

Trees reach up for sunlight. Tall trees—with their high, lush canopies of leaves—shade out smaller trees. The roots of some tropical plants produce chemicals that prevent rivals from growing nearby. Others have leaves or bark or sap that poison insects. On some trees, nasty spines keep leaf eaters away. Small toads, colored bright red, yellow, or blue, ooze poisons. Some plants grow high above the ground on tree branches or leaves.

Swarms of insects hum and buzz. Monkeys howl, birds scream and whistle. Snakes slither along the ground or race through the foliage. Eagles and hawks glide overhead. Cats prowl below.

Such rain forests cover just 2 percent of the world, but they contain more than half of the world's plant and animal species. However, growing human populations are cutting the forests to create lumber and farmlands. Every year, people cut an amount of rain forest equal in area to England, Scotland, and Wales.

Conservationists are trying to slow the destruction. For instance, they are helping to create rain forest preserves where local people can harvest nuts, fruits, rubber, cocoa, medicines, and other valuable products without hurting the forests. Environmental organizations also work with governments to create rain forest national parks.

The conservationists are racing against time. If the destruction is not slowed, the rain forests will be gone in about 50 years.

10

Humpback Whales Are Still Singing

Deep in the Atlantic Ocean, a humpback whale is singing. It is a strange song: low moans and high trills. It goes on and on.

No human knows what this song of the humpback whale is about. But human scientists are dropping microphones into the ocean to eavesdrop on the humpbacks' singing, trying to understand.

Humpbacks are as big as city buses. Yet, their songs are delicate and full of complexity. Scientists who study the whale songs have found they go on for hours. A scientist once recorded a whale that sang without stopping for 21 hours. Whales cover every note humans can hear, high to low. But researchers are still unsure why whales put so much energy into singing their underwater solos.

Mainly, the humpbacks sing in the fall and winter, when they migrate to the tropics to mate and raise their young, or *calves*. Atlantic humpbacks go to the Caribbean. Pacific humpbacks winter around the Hawaiian Islands and off the west coast of Mexico. In the summer, humpbacks swim north to colder waters, where they rarely sing. But their tropical wintering waters are full of whale song.

Only males sing. Biologists suspect the songs are involved in whale mating, but nobody knows for sure. What scientists do know is that a whale song is not like a bird song. Most bird species have only one song, always the same. But humpbacks *change* their songs.

Over vast stretches of the Atlantic or Pacific, all the whales sing the same song. But as weeks go by, the songs slowly change—a note here, a group of notes there. Eventually, the song is altogether new and the old song is never heard again. In addition, the whales often sing in rhyme, just as the words of human songs or poems often regularly repeat certain sounds. Scientists suspect rhyme helps the humpbacks remember long songs.

Other whale species also sing and make sounds. For instance, a Cornell University researcher, Chris Clark, has cut holes in the

HUMPBACK WHALE

Arctic ice off Alaska to drop microphones into the ocean. From his camp on the creaking and crunching ice, he listens to the calls of bowhead whales migrating from their wintering grounds in the Bering Sea to their summer homes in Canada's Beaufort Sea. He has found that bowheads make underwater noises for a number of reasons.

For one thing, the whales listen to their own sounds and the sounds of their bowhead friends. They also listen to the echoes as their moans and cries bounce off the bottom and off ice floating overhead. From the sounds and their echoes, the whales can picture their watery world. Without seeing far in the murky, dark water, they know where the bottom forms a hill ahead of them and where an iceberg above extends down deep into the water. Then they can swim around such obstacles.

Also, by exchanging calls and cries, the whales keep track of each other. That way, 10 whales spread out over 10 square miles of ocean can all swim together in one direction.

By listening to bowhead whales, Chris Clark discovered there are more bowhead whales than scientists realized. When Clark began his studies in 1979, biologists thought only about 1,500 bowheads were left alive in the world. But Clark counted about 3,000. Yet, that was far fewer than the 20,000 bowheads that swam in Arctic waters before whale hunters nearly exterminated them.

Of the 32 species of whales and dolphins around the world, eight species of whale and two species of dolphin (smaller cousins of the whale) are listed as endangered. The endangered whales include the bowhead and the humpback. Another endangered whale is the blue whale, the largest animal on earth. Blue whales grow to 100 feet, longer than a basketball court.

Whales have fins instead of legs, and they are streamlined for swimming in the oceans. But they are mammals, just like bears, dogs, and us. In fact, millions of years ago, ancestors of modern whales lived on land and walked on legs. Gradually, they returned to the oceans and evolved fins. However, whales still have lungs, just as we do, and they must come to the surface to breathe. But they can hold their breath a lot longer than their human cousins. The sperm whale can hold its breath for two hours and swim more than a mile deep.

Most whales have teeth. They eat fish, squid, octopuses, and other sea creatures. Some whales, like the humpback, have no teeth. Instead, hanging from the roofs of their mouths are thin plates of bone, called *baleen*. The baleen enables the humpbacks to strain tiny sea animals, called *plankton*, out of the water. The humpbacks swim with their huge mouths open, collecting the plankton on their baleen. When they have a mouthful, they lick the plankton off the baleen with their tongue.

Whalers hunted humpbacks and other species nearly to extinction to get their baleen, or "whalebone," which once was used in women's clothes. Whalers also killed whales by the millions to get their thick fat, called *blubber*, which could be boiled down to an oil used in lamps. The whalers also killed whales for their meat, which is eaten in some countries.

Many people are opposed to whale hunting because researchers believe whales are among the most intelligent animals. Also, whales are gentle creatures that cause no harm. And mother whales are devoted to their young. As species after species of whale began to dwindle away, many people called for an end to the whaling industry, in which hunters in large boats pursue whales and shoot spearlike harpoons into them. As a result, since 1946, the International Whaling Commission has regulated the number of whales of the different species that hunters can kill each year. So far, the controls on whaling seem to be helping at least some species.

The major whaling nations—Japan, Norway, and Iceland—have been fighting limits on whale hunting. Whale supporters have been urging the nations of the world to ban forever the killing of whales. They argue that whales belong to the Earth itself, not to hunters armed with exploding harpoons who kill these huge mammal cousins of ours to make money by selling their meat.

On land, the argument goes on. Deep in the Atlantic and the Pacific, humpback whales are swimming. And, as they swim, they sing. Their songs are long, and humans cannot yet understand them. But many hope the whales will be able to sing on forever.

11

The Man Who Walks for Rhinoceroses

In Kenya, Africa, an ordinary citizen named Michael Werikhe has devoted himself to saving rhinoceroses.

He began in 1982, when he walked 300 miles from his hometown of Mombasa to the big city of Nairobi. He told everyone he met along the way that rhinoceroses are endangered. For every mile he walked, he asked people to donate money to help protect rhinoceroses.

Since then, Werikhe (pronounced Where-ree-key) has walked 3,400 miles across Africa and Europe, raising interest in rhinoceroses. In 1991, he walked 1,500 miles through the United States and Canada to raise more funds for rhinoceros protection.

"It is a bit of a sacrifice, because personally I don't like walking long distances," Werikhe has said, with a smile. "It's my way of saying thank you to nature and it's my little contribution."

His interest in endangered species began when he was a child. As a boy, he brought home injured animals and cared for them until they could return to the wild. Later, he took a job sorting piles of elephant tusks and rhinoceros horns, before it became illegal to hunt the animals. Seeing the evidence of so many animals killed made him so sad that he quit the job. Now he works for an industrial organization in Mombasa.

"I made a decision—or rather a promise to myself—to do something for nature, something to save the rhinos," he has told reporters. "I am a very simple man," he said, explaining his simple plan to save the animals.

His walks may be simple, but they have been successful. In Europe alone his walk raised $1 million for rhinoceros conservation.

Rhinoceroses are worth saving. Of all the mammals that ever lived on land, the largest was a rhinoceros. It could have stood in your backyard and looked into an upstairs window.

That huge rhinoceros disappeared

WHITE RHINOCEROS

7 million years ago, when an Ice Age chilled the Earth. It was one of over 100 rhinoceros species that once roamed the planet's grasslands and forests. Some of the extinct rhinoceroses were huge. Some were no bigger than collie dogs. Some were fat wallowers in ponds and others were skinny runners. Some had woolly fur.

Today, only five rhinoceros species are left: three in Asia and two in Africa. If humans do not stop shooting them, soon no rhinoceroses will be left at all.

Rhinoceroses are relatives of the horse. All five modern species are stocky and muscular, with horns on their snouts. Their skin has thick folds, like armor plating. They have blurry vision, but sharp ears and noses. They spend their days placidly munching grasses and other plants. However, if they lose their temper, they may charge, even at a human. A rhinoceros can run 30 miles per hour and it weighs up to five tons. Being charged by a rhinoceros is like being attacked by a pick-up truck.

Modern rhinoceroses are almost hairless, except for their horns. The horns are made of pressed-together hair. Rhinoceroses use their horns to defend themselves and sometimes to plow up the ground looking for roots to eat. But it is mainly because of their horns that rhinoceroses are endangered.

In Arab countries, daggers with rhinoceros-horn handles are highly valued. In Asia, powdered rhinoceros horn is considered a cure for fevers. The horn is so valuable that it sells for up to $25,000 per pound. All rhinoceroses are endangered and all are protected by law. However, hunters shoot rhinoceroses illegally, which is called *poaching*. After they shoot a rhinoceros, the poachers cut off its horn to sell.

The shooting has pushed the rhinoceros close to extinction. Today, only about 11,000 rhinoceroses of all five species are alive. Now governments and wildlife experts are trying to save the rhinoceroses.

One step is to increase the number of armed wardens who guard rhinoceroses. In 1974, the small Asian country of Nepal sent hundreds of soldiers to guard the national park where its rhinoceroses lived. At the time, Nepal had only about 95 rhinoceroses left alive. Today the park's rhinoceroses number approximately 400, with two soldiers to guard each rhinoceros.

In Namibia, Africa, government wardens inject rhinoceroses with medicines that briefly put them to sleep. Then the wardens cut off the rhinoceroses' horns. That way, poachers have no reason to shoot the rhinoceroses. The horns grow back. Meanwhile, the animals have no difficulty defending themselves and finding food, even without their horns. Officials also are persuading African and Asian villagers that rhinoceroses are more valuable to them alive because the animals attract tourists.

The number of rhinoceroses continues to

decline. One reason is that several Asian countries still import rhinoceros horn for medicines.

Efforts to save rhinoceroses have had *some* effect: they are not disappearing as fast as they were just a few years ago. But they will continue to drift toward extinction until the poaching is stopped completely.

"There's no way man can survive on this planet alone," says Michael Werikhe, the man who walks for rhinoceroses. "We have to protect nature as well as protect man."

12

Sea Otters Are Swimming Back

A newborn sea otter pup floats alone in the Pacific, crying for her mother. But her mother is gone.

The pup was a twin, and mother otters can care for only one pup at a time. This mother otter clasped the first of her newborns to her chest and swam away. Now the second pup floats alone, off the California coast. Special baby fur keeps her afloat like a cork. But, alone, she will starve.

She is lucky, a wildlife official spots her. He rushes the orphaned pup to California's Monterey Bay Aquarium, which specializes in caring for orphaned otters. But the pup—one day old and only 2 ½ pounds—is the tiniest otter the aquarium staff has ever seen.

Volunteer otter-sitters and aquarium workers care for her around the clock. They brush and comb her. From syringes, they feed her an otter "milkshake," a mixture of clams, squid, fish oil, milk, vitamins, and minerals. Because the pup repeatedly spills her food on herself, she is named "Milkdud."

Milkdud sleeps on a waterbed. She learns to swim in a saltwater pool, and she plays with rubber balls and frisbees. An older aquarium otter, Hailey, becomes Milkdud's substitute mother, grooming and protecting her. Hailey even lets Milkdud steal food from her. As Milkdud grows, she moves to a large outdoor tank stocked with kelp, snails, crabs, and sea urchins.

Today Milkdud is one of four sea otters exhibited at the aquarium. All orphans, they cannot be released to the wild because they could not survive on their own. But now the Monterey Bay Aquarium has developed a method for preparing otter pups for a return to the ocean. A human "mother" swims with the young otters in the ocean by the aquarium, helping them learn to find food and adjust to the wild. The aquarium has successfully returned several otter pups to the Pacific. Its goal is to help bring back the southern sea otter, an animal that almost disappeared.

In 1741, Russian explorers discovered sea otters in Alaska. Within 50 years, fur traders had nearly exterminated Alaskan sea otters, hunting them for their luxurious coats. The traders moved south to California, killing otters up and down the coast. By the early 1900s, everyone thought the California sea otter had been hunted to extinction.

However, in 1915, scientists saw otters in an isolated California inlet. Laws now protected otters. But the scientists feared uncaring people would ignore the laws, so they kept their sighting of the otters secret.

Today, California sea otters are making a slow comeback. Over 2,000 otters now live in colonies along the central California coast.

Scientists studying California sea otters find they spend much of their time near the shore in kelp, a marine plant that grows from the ocean floor. Kelp forms dense forests in the ocean. To rest, otters drape themselves in kelp strands to keep from drifting. Otters often rest in groups, called *rafts*.

California sea otters are about the size of basset hounds. They have whiskered snouts, stubby front paws, and webbed hind flippers that look like baseball mitts. Whales and seals—also sea mammals—keep warm with layers of blubber. But the sea otter relies instead on its thick, water-resistant fur. Otters

SEA OTTER

actually have two layers of fur. One is an outer layer of coarse guard hairs. These protect and waterproof a second, inner layer of dense hair called *underfur*. This underfur has up to one million hairs *per square inch*. A dog has only 1,000 to 60,000 hairs per square inch.

The sea otter spends a lot of time grooming and cleaning its thick fur. It vigorously rubs its fur with its forepaws and then somersaults and barrel-rolls in the water. This enables the otter to trap air bubbles in its fur. These air bubbles keep the otter warm and dry. Otters also have special oil glands that help make their fur water repellant.

An otter's paws and hind flippers are hairless. To keep them warm, otters often swim on their backs, holding their paws and flippers up in the air. Some rest their front paws on their eyes.

Otters are among the few mammals that use tools. As they swim on their backs, they keep a flat stone on their bellies. They bang clams, abalones, mussels, and other shellfish against the stone to crack the shells. Then they eat the meat inside. They also eat sea urchins and crabs. One female wild otter in Monterey Bay carries a glass cola bottle on her stomach. She found the discarded bottle in the water and uses it to pry shellfish out of rock crevices. Now her daughter uses a cola bottle, just like Mom!

Female otters are devoted mothers. A pup spends most of its time riding on its mother's belly, learning the complex skills it needs to survive. A pup learning to eat will pretend it has a rock on its belly and wave the food in the air, imitating the pounding motions its mother makes to crack open shellfish. Because of the special natal fur that keeps it afloat, a pup cannot dive for the first three months of its life. Instead, it sticks its head underwater, like a snorkeler, and watches its mother forage for food. Male otters sometimes snatch pups from their mothers. The male holds the pup hostage until its mother ransoms the pup by giving the male food.

Otters eat up to eight pounds of food a day. For a person, that would be like eating 140 hamburgers every day. Otters keep warm by burning calories rapidly. In fact, sometimes otters eat so many shellfish, like abalone, that commercial fisherman are annoyed.

Oil spills threaten otters. The spilled oil quickly mats an otter's fur, destroying its insulation. Within hours, the otter freezes to death. When the otter tries to lick its soiled fur, it ingests the toxic oil. Hundreds of Alaskan otters died after the tanker *Exxon Valdez* spilled over 10 million gallons of crude oil into Alaska's Prince William Sound.

Protected, California sea otters are making a comeback. Scientists who study them believe the otters may survive.

13

Success Story: The Northern Beaver

Beavers teach an important lesson: Left alone, even some nearly extinct species bounce back.

Beavers are 60-pound rodents, cousins of mice, woodchucks, and squirrels, that live in water. With their big front teeth, they eat trees that live beside streams and ponds. Beavers weave together twigs and branches to dam up streams, using mud as cement. New Hampshire beavers once built a dam three-quarters of a mile long.

In the pond that forms behind their dam, a beaver couple builds a domed house of twigs and mud. Here they raise their young, which takes two years. The house also keeps them warm and dry in the winter. Its entrance is in the floor, so that the only way to get inside is by swimming underwater. That keeps out beaver-eating predators, like foxes and coyotes.

Beavers gnaw down trees growing along the shore. A Canadian beaver once gnawed down a cottonwood tree that stood 110 feet tall and 5 feet thick. Usually, however, beavers cut down smaller trees. They tow the trees into the pond and fasten them underwater to the bottom. All winter, when the pond is iced over, the drowned trees serve as beaver food.

Because it lives in cold water, the beaver's fur is thick. That fur was almost its downfall.

Before Europeans came to North America, beavers numbered in the millions. But fancy dressers in Europe and America liked to wear beaver-skin hats and coats. By the year 1900, beavers had been trapped nearly into extinction, turned into hats and coats.

After that, however, fashions changed. For some reason, people no longer wanted beaver coats and hats. People ignored the few beavers remaining. Left alone, the animals multiplied. Today the United States has nearly 2 million beavers. Canada has many more.

Now humans are discovering that beavers can be more useful alive than turned into hats. From New York to Wyoming, wildlife

BEAVER

officials are transplanting beavers back into streams where they once lived. That is because the ponds behind beaver dams are good habitat—or homes—for fish and other water creatures, from ducks and herons to mink. And the ponds have other advantages, too.

Researchers have shown that beaver ponds change the vegetation around a stream, creating more food for grazing animals such as deer. The ponds also help stop land from eroding in areas where cattle have grazed too heavily.

Unfortunately, beavers see no difference between a natural stream and a stream flowing through a culvert, where the beaver dam might flood somebody's lawn. And they may find a shade tree growing on the lawn as tasty as a wild aspen.

Still, the beaver now has many friends. It looks as if beaver dams have returned for good to the continent's streams.

14

The New Arks

In a Bible story, God becomes so angry with humans that He decides to flood the Earth and destroy His creations. But He finds one good person, Noah, and God decides against extinction.

God tells Noah to build a huge ark, or boat. And He tells Noah to collect one male and one female of each species and put them in the ark. Noah builds the ark and fills it with doves, giraffes, dandelions, houseflies, and every other living thing, a male and female of each species.

It rains 40 days and 40 nights, flooding the Earth. Everything drowns . . . except the creatures in the ark.

Finally the rain stops and the flood drains away. Noah and his family leave the ark. So do all the birds, animals, insects, plants, and other living things Noah collected. Each species goes back into the world to mate and start again.

Today, the world faces a new sort of flood—a "flood" of human beings. Humans are reproducing so rapidly that scientists expect our numbers to double within about 50 years. We are using up more and more land and creating ever more pollution. In the process, we destroy "habitat"—the prairies, marshes, forests, deserts, streams, lakes, and oceans where animals and plants live.

From the year one until the 1600s, when the Pilgrims arrived in Massachusetts, one or two species a year became extinct. Now, species are disappearing by the *hour*. Some researchers believe a million species will become extinct in just the next 20 years.

Many scientists believe our numbers are growing too fast for our own good. If humans do nothing to stop our population from growing so fast, they believe that in about 200 years we will no longer be able to live on the Earth in such numbers. Then millions may die in disasters. Many may starve as we run out of land for farming. Epidemics may kill millions

PRZEWALSKI'S HORSE

more. So may wars, as crowded humans fight each other for food, water, and other resources.

If all that comes to pass, in about 200 years the human "flood" may recede. Once again, habitat will be available for wildlife. However, by then, thousands of species may be extinct. To try to save at least some species that would otherwise disappear, zoos and aquariums have begun to act as modern arks.

A pioneer of zoos-as-arks was the British naturalist Gerald Durrell. He had loved animals since he was small child; if he was not taken to visit a nearby zoo every day, he threw tantrums. In 1959, on the tiny island of Jersey in the English Channel, he created a new kind of zoo, where animals live in more natural settings. At Durrell's zoo, endangered species would breed in the protection of captivity, and their offspring would help keep the species alive. Eventually, his work led to an international organization called the Wildlife Preservation Trust. Other zoos around the world also began breeding endangered species.

In 1980, many zoos and aquariums began working together to keep at least some endangered species alive for the next 200 years. They hope that, 200 years from now, habitat may again be available so that the species bred in captivity can be returned to the wild.

First, the zoos began making the exhibits where zoo animals live as natural as possible. Zoo staff plant the right species of trees, shrubs, grasses, and other vegetation in the exhibits. For species that naturally live in tropical forests, they create artificial rains. Other species, like otters, may have an artificial stream for swimming. Mountain goats may have artificial cliffs to climb. The aim is to keep the animals as wild as possible. Also, by placing animals in their normal surroundings, scientists can better study their behavior.

Besides making their exhibits more natural, the zoos also are learning to breed the animals in captivity. In fact, about 90 percent of the mammals and 75 percent of the birds now exhibited in zoos were born at the zoo.

That way, zoos are not robbing the wild of rare animals. Also, if a species becomes extinct in the wild, it may live on in zoos until it can return to its natural habitat.

Some researchers are now learning to "grow" such habitats as marshes, prairies, or tropical forests where zoos can release captive-bred species. Zoos already have had some success returning animals to the wild.

For instance, seven zoos in the United States and Canada have been breeding the crested toad, found only in Puerto Rico. One variety of crested toad is down to just 25 individuals in the wild. Now the zoos have raised about 600 crested-toad tadpoles in captivity and they are beginning to return the toads to Puerto Rico.

Other zoos in Britain and the United States are preparing to return a wild horse to the Gobi Desert of Mongolia. Called Przewalski's horse after the Polish explorer who discovered it, the horse has been extinct in the wild since the 1960s. However, nearly 1,000 of the rare horses live and breed in zoos around the world. Now the zoos are working with Mongolian scientists to return the horses to their native desert. Other zoos are returning a species of antelope, the Arabian oryx, to its native Middle Eastern deserts. It was hunted to extinction in the wild.

However, the zoos say their efforts can be only a small part of the effort to save species. Even if zoos manage to keep some species alive in captivity, it will be of little use if no wild places are left to which the animals can return.

The only real solution to vanishing species is to save large areas of tropical forest, grassland, marsh, northern forest, and other habitats from destruction. That is difficult and sometimes impossible. A growing human population cuts trees for logs, builds houses and factories, and turns forests into farmland and cattle grazing land.

Also, zoos are simply too few and too small to save more than a fraction of the endangered species. One zoo official estimates that some 2,000 species of large animals alone will become extinct over the next two decades. But zoos have space to breed only about 900 of those species.

Captive-breeding programs face other difficulties, too. Animals bred in captivity may no longer be able to survive in the wild, for instance. And returning even one or two animals to the wild can be extremely expensive. Some people believe it is wrong to keep *any* animals in captivity.

Yet, many scientists working with endangered species approve the zoos' efforts to breed endangered animals in captivity. Noah's ark saved *all* species. Zoos may save only a few. But a few, the scientists argue, is better than none at all.

Consider the Mauritius kestrel, a small falcon that lives only on the Indian Ocean island of Mauritius. By the late 1970s, the island's forests had been chopped down and only six of the kestrels were left alive. Then a European zoo captured a few and bred them in captivity. Meanwhile, the Mauritius government turned the island's last bit of forest land into a kestrel preserve. Since 1985, the zoo has released more than 100 of the kestrels back into the wild.

Mauritius was once the home of the dodo. Human settlers drove these large birds to extinction. No dodos are alive today. However, because of a zoo's captive-breeding program, the Mauritius kestrel is once again flying over the island.

Birds: Dinosaurs with Feathers

Many scientists believe birds are the dinosaurs' closest living relatives. About 150 million years ago, certain dinosaurs evolved into birds. The scales of the dinosaurs changed into feathers.

Ornithologists—scientists who study birds—know of approximately 8,500 bird species, worldwide. All have feathers and all reproduce by laying eggs. Some birds, like penguins and ostriches, have lost the ability to fly. But their ancestors flew, for birds are designed as creatures of the air.

For one thing, birds' forelimbs (the equivalent of our arms) have evolved into wings. Also, to make flying easier, birds are lightweight. Their bones are hollow and their beaks weigh little. Birds are streamlined to slide easily through the air. Their necks are highly flexible, so they can look all around for danger or prey. Also, they can twist their heads to preen their feathers, cleaning them and spreading oil through them. This helps the birds stay warm and dry.

Like mammals, birds are warm-blooded. But birds' bodies are much warmer and their hearts beat faster. Human hearts beat about 70 times a minute. A hummingbird's heart beats 615 times a minute.

Birds are like mammals in another important way: they also care for their young. Birds feed and protect their chicks until they are ready to fly off on their own.

15

The Condor Returns to the Wild

AC-9 edged toward the dead calf. It lay in California's Los Padres National Forest. Nothing dangerous seemed to stir. But AC-9 was wary.

Wings spanning 10 feet, he was baron of the skies, his feathers black, his head naked. But he was awkward on the ground. He took another step toward the dead calf.

Boom!

He tried to fly, but the net shot from a cannon wrapped around him. AC-9—Adult Condor-9—was captured, the last wild California condor.

That was in April, 1987. For the first time since the Ice Age, when giant beavers and woolly mammoths stalked the continent, no condors flew over North America.

But the hiding man who shot the net over AC-9 was a friend. He was a biologist, part of an effort to save the California condor from extinction.

Condors are large vultures, ancient birds that clean away the carcasses of animals that die. Eleven thousand years ago, they soared across the West and as far east as what is now Florida and New York. But as the Ice Age ended and the climate warmed, many of the epoch's huge beasts became extinct. That meant fewer carcasses for the condors to find. They, too, began to disappear. During the California Gold Rush of 1849, miners shot condors for "fun." Later, pesticides sickened them. And condors feeding on carcasses left by hunters were poisoned by lead shot.

It takes six years for a condor to mature enough to reproduce, and wild condors raise only one chick every two years. With such a slow rate of reproduction, the big birds could not bounce back from the shootings and poisonings. By the 1980s, the species was down to just a few individuals. Like AC-9, each condor had a name supplied by biologists.

Scientists tried to save the birds by

CALIFORNIA CONDOR

stealing their eggs. It sounds destructive, but the biologists knew a secret: if a condor pair's egg is taken just after it is laid, they will quickly produce a new egg. And so by stealing condor eggs at just the right moment, the scientists prodded the condors into producing more eggs. Meanwhile, they hatched the stolen eggs and raised the chicks in captivity. That way, even if the condors disappeared in the wild, at least the species would still exist in zoos. And in 1985 it looked as if the condors were disappearing: of the 15 wild condors left, 9 disappeared.

Scientists believed that with only seven condors left in the wild, they had to act. And so they captured the last wild condors and placed them in protective captivity at the Los Angeles Zoo and the San Diego Wild Animal Park. The captured wild condors and condors already in the zoos totalled only 27, all the California condors left in the world.

Capturing the last of the condors was controversial. Many environmentalists argued that the birds should be left in the wild. But a year after AC-9's capture, a female condor chick hatched at the San Diego Wild Animal Park. Biologists named her "Molloko."

More condor chicks hatched in the zoos. By the summer of 1991, the number of California condors had increased from 27 to 52 birds. Scientists were planning to return condors to the wild.

In the Los Padres National Forest about an hour's drive northwest of Los Angeles, they created a condor sanctuary. In a cave high on a mountainside in the sanctuary, they raised four Andean condor chicks—South American cousins of the California condor—and turned them loose. The aim was to see if condors raised in captivity could successfully return to the wild. It also was to see if condors could be trained to return to their cave to feed on meat provided by the biologists. Otherwise, the scientists feared, freed condors would eat lead shot from animals killed by hunters and die.

The experiment with the Andean condors was a success. And so, just after Christmas, 1991, scientists released into the wild two California condors hatched and raised in captivity.

Biologists hope to release more condors to help the species begin again. But the world has changed since the Ice Age. Now millions of people crowd California, with cars, power lines, and oil rigs. Nobody knows if the condors can adjust.

Still, in 1991, with help from humans, condors again flew wild over California.

16

American Eagles Are Soaring Again

Ross and his mate—American bald eagles—were off hunting mice for their chicks. That was when biologists climbed their tree to check their nest.

Ross was raised by humans, but now he is wild.

A decade ago, in all states except Alaska, the bald eagle was virtually extinct. Then biologists began a nationwide effort to raise eagles in captivity and return them to the wild. Ross, as biologists named him, was the first eagle released in Massachusetts.

Ross mated with another released captive. In 1989, the pair produced Massachusetts' first wild-born eagle chicks in nearly a century. In 1990, they lost their own chick to parasites, but successfully reared a foster eaglet, hatched in captivity and placed in their nest by biologists. Now it was a new

spring, time to check the pair's nest once again.

Biologists clambered 55 feet up a red oak to the nest, a mound of sticks four feet across. Eagles reuse their nests year after year. This nest overlooked the forested mountains of western Massachusetts and blue Quabbin Reservoir.

Inside the nest were two eaglets. Their yellow eyes and hooked beaks already looked fierce. Gently, the biologists stuffed the chicks into cloth bags and lowered them to the ground. There they weighed the chicks and checked their ears, eyes, and throats. Satisfied that the eaglets were healthy, they returned them to their nest and slipped away into the forest.

Eagles, helped by humans, are slowly returning. But it was humans who nearly killed off these great hunting birds. With their broad, dark-brown wings, their snow-white heads and tails, eagles look invulnerable. Yet, a chemical called DDT brought them down.

People sprayed farm crops and trees with DDT to destroy plant-eating insects and mosquitoes. But DDT accumulated in the soil and water, getting into the small animals that eagles and other predatory birds eat. The pesticide built up in the eagles' bodies. One result, biologists discovered, was that the shells of the eagles' eggs became so thin that they broke. No eaglets could hatch. Once eagles lived from Florida to Maine, along every lake and river, and from southern California to Alaska in the west. But by the time officials banned DDT in 1972, the eagle—our national symbol—was almost gone in the 48 contiguous U.S. states.

In 1980 only about 650 pairs of bald eagles remained in the lower 48 states. The nation began an effort to repopulate the country with eagles. State and federal agencies and civilian volunteers are all taking part.

Arizona, for instance, began a "Nest Watch" program, with volunteers guarding eagle chicks around the clock. If the chicks fell from their nests, the volunteers rescued them. They also chased off bobcats, rock climbers, canoers, or bird watchers who came too close. One volunteer shinnied up a tree to take out a fishing lure stuck in one bird's beak.

Efforts like those in Massachusetts and Arizona have seen the bald eagle count climb to about 8,000. But obstacles remain.

BALD EAGLE

In Oklahoma, a bald eagle recently died in an open oil pit. Such uncovered pits created by the oil industry in Texas, Oklahoma, New Mexico, Kansas, and Colorado look like ponds to thirsty migrating birds. When the birds land in the oil, they die. In just five states, uncovered oil pits kill 500,000 ducks every year. Now they have killed at least one eagle.

Meanwhile, not long ago, a biologist made a carefully timed raid on an eagle's nest in northern Florida. He knew that if a mother eagle lost an egg at that moment in her reproductive cycle, she would quickly lay a new egg. The biologist rushed the captured egg to a wildlife farm in Oklahoma and put it under a nesting chicken. When the egg hatched, biologists carefully nurtured the eaglet to adulthood, naming her Eagle A-40. They released her to the wild near the Mississippi coast and cheered as A-40 soared away.

Months later, a passerby found an eagle in southern Georgia, her feathers spattered with blood. It was A-40. She had been shot, one of three eagles shot in Georgia in just three months.

Some humans are struggling to return bald eagles to the wild. Yet, other humans are shooting eagles. Wildlife officials believe some of the shooters are hunters excited by the chance to kill so magnificent a bird, even though it is illegal. And some farmers kill eagles because they believe the birds will attack their chickens, even though eagles rarely prey on domestic fowl. Meanwhile, eagles also are struck by automobiles. And developers and loggers cut down the tall pines beside ponds, lakes, and rivers where eagles nest. In Maine, wildlife officials worry that DDT sprayed on forests and farms before 1963 is still affecting eagles because they have not been reproducing as rapidly as biologists hoped.

Even so, across the country, eagles are slowly coming back. All Americans may yet be able to see eagles soaring.

17

Peregrine Falcons Find New Homes

Every spring, people gather on East River Drive in Manhattan to stare through binoculars at New York Hospital. They are watching something extraordinary—birds.

On a ledge 25 stories up from the street, a pair of peregrine falcons nest. They return year after year. To them, the hospital is a cliff. Their kind has always nested on mountain cliffs, and the country's largest city, to these wildest of birds, is a complex of canyons. For them, the city is rich in game, especially pigeons and starlings. But what is extraordinary about the peregrines is not that they live in New York City. It is that they live at all.

By the 1960s, biologists realized that peregrine falcons were vanishing because of the widespread use of the pesticide DDT. As it did to bald eagles, DDT made the shells of the falcon's eggs too thin. Unable to hatch their

chicks, peregrines had vanished all across America. Fewer than 50 pairs were left, all in the Rocky Mountains. For people who loved wildlife, it was a terrible loss.

Since ancient times, humans have admired peregrine falcons. As one scientist wrote in 1904: "The Peregrine Falcon is, perhaps, the most highly specialized and superlatively well developed flying organism on our planet today."

Peregrines are living jet fighters. Their twists and turns in the sky amaze the eye, and so do their power dives: a peregrine swoops down on its prey at 200 miles per hour. The government banned the use of DDT and related chemicals. But for the peregrines it seemed too late.

One ornithologist, Dr. Tom Cade, at Cornell University, refused to let the peregrines become extinct. In 1970, he began an effort to raise peregrines in captivity. In a barn near the university, he painstakingly learned to breed the birds and nurture their brown-and-white eggs until they hatched. Soon the project expanded to include a second breeding facility, in Colorado. When enough birds were bred in the two barns and carefully raised to

remain wild, biologists began releasing them across the country.

Today, some 1,200 peregrine falcons are living wild again. They are still threatened by pesticides in some areas, such as California. But, overall, their numbers are steadily growing. In rural states, like Vermont, they are nesting once more on high mountainside cliffs. But the falcons also are colonizing cities like Boston, New York, and Baltimore. Biologists are finding that urban life agrees with these birds of prey: city peregrines are reproducing faster than their country cousins.

DDT is gone. In a rare success for endangered wildlife, peregrine falcons are back. And they seem to have a future in America's skies.

18

War Brought Peace to the Cranes of Korea

In the 1950s, the Korean peninsula was at war. Oddly, the fierce fighting may have saved two of Asia's endangered cranes.

South Korea, backed by United States and the United Nations, battled North Korea and China. Thousands on both sides were hurt or killed. The bombing and shelling devastated the land. When the terrible fighting finally ended in 1953, North Korea was separated from South Korea by a fenced-off strip of land. It was called the Demilitarized Zone, or DMZ. No people could go there. For cranes, the DMZ has been a haven.

Cranes are long-lived and mate for life. Throughout Asia they are considered symbols of longevity, happy marriages, and good luck.

But without the DMZ, the cranes were running out of habitat.

The DMZ cuts across the Korean peninsula. It is 2.5 miles wide and 151 miles long. The rest of Korea is thick with farms, villages, and cities, but nobody lives in the mountains of the DMZ. Armed soldiers on both sides keep people out. As a result, in the decades since the war, the DMZ has become a wild preserve of marshes, meadows, and forests of oaks, pines, and maples, where lynx and tigers prowl. Here and there a bombed bridge rusts or a shelled farmhouse lies buried in pussy willows.

"It is one of the world's great wildlife paradises," says ornithologist George Archibald, who heads the International Crane Foundation, in Baraboo, Wisconsin.

Archibald first visited the DMZ in 1974 to study cranes—large, long-legged wading birds. Relations between North and South Korea were tense, and mortar bursts punctuated the nights. But Archibald was delighted with what he saw. From his camp with a South Korean army unit, Archibald looked down on Asia's white-naped cranes, one of the world's eight endangered crane species.

JAPANESE CRANE

White-naped cranes have declined because of hunting and habitat loss. They wade through salt marshes on their stiltlike legs, poking their long beaks into the mud to dig up grass seeds and plant roots. Their appearance is striking. They have big red eye patches and pure white feathers down the back of their necks. Their wings are light gray and their chests are dark gray.

During the night, most of the 2,000 or so white-naped cranes that winter in the DMZ roost together in marshes and mudflats. Some of the cranes live as pairs, caring for one or two chicks. Such families claim a patch of marsh as "theirs," trumpeting warnings at other cranes that trespass. After a huge flock of cranes has fed on a mudflat, poking it with their bills, it looks as if farmers have worked the soil with hoes. Archibald's research suggests that the DMZ has been good to white-napeds. Their numbers seem to be increasing.

For another endangered species, Japanese cranes, the DMZ may be even more important. These white birds are as tall as a human, with black rumps and bald, red heads. Only about 400 Japanese cranes are left in the world, and 200 of them winter in the Korean DMZ. There they wade through the water, hunting worms, leeches, and other small marsh animals. On frigid days, they huddle together beside the rivers, hunting little and living on the fat accumulated in their bodies.

In the spring, the Japanese cranes fill the air with their trumpeting as they flap to Asia's northern wetlands. There they perform their strange courtship dances to attract mates. Once a pair has mated and the female has laid her eggs, the male sits on the eggs all day and the female sits on them all night. After a month, the chicks hatch. No bigger than a human fist, they toddle after their parents into the marsh to hunt for food, beginning a spectacular growth. Within just three months the tiny chick grows to five feet tall, with wings that span eight feet.

Unlike the white-naped cranes, which are increasing, the Japanese cranes are merely holding steady. George Archibald believes it is because human development is destroying the wetlands in northern China where the cranes breed. Without their haven in the DMZ, the Japanese cranes might already have disappeared.

Recently, North and South Korea have been negotiating a reunification of the peninsula into a single country. That would be a major step toward peace in the region. However, it might also mean people could once again return to the DMZ, a disaster for the cranes. George Archibald and Korean ornithologists are working to keep the DMZ as a haven for wildlife. Perhaps they will succeed.

19

Northern Spotted Owls: Unwilling Celebrities

Eyes big, the northern spotted owl sits high in a 500-year-old Douglas fir in Oregon's Cascade Mountains. It turns its head almost backwards, locating a sound.

Sometimes when it listens this way, the owl hears chain saws whine—loggers cutting down the old trees to haul to lumber mills. This time, however, the owl hears four hoots, a call like its own. It flies to investigate.

The owl flaps silently past huge tree trunks that rise like columns to a roof of needles and branches. It is so dark under the trees that few bushes grow to impede the owl's flight. Here and there, a shaft of sunlight slants down. In one lighted patch the owl finds a federal biologist, who has been imitating its call. He wears a down vest against the mountain air, which is cool and damp. In his gloved hand he holds a white laboratory mouse.

The owl swoops, seizing the mouse in its talons. As the owl flaps away, the biologist follows. The owl did not eat the mouse immediately: that means it is bringing the prey back to its young. The biologist will follow the owl to its nest and count the chicks. But he will keep the nest's location a secret, fearing that somebody might shoot the endangered owls.

Northern spotted owls have many friends these days. But they also have angry enemies. Nobody objects to the bird itself, for the owl is harmless to all except mice. And mice are plentiful. But even the owl's friends see it as part of a much larger issue: the survival of ancient forests.

Environmentalists and loggers are fighting over the last of ancient coastal forests of northern California, Oregon, and Washington. The northern spotted owl is a draftee on the environmental side. That is because it is disappearing along with its ancient forest habitat. Only about 5,000 owls remain.

It is a sad war. A species' life is at stake, but so is the livelihood of loggers. It is a

NORTHERN SPOTTED OWL

complicated war, because the small fraction of the West's ancient forests that remain are virtually all on public lands, owned by the taxpayers and managed by the U.S. Forest Service.

In the Pacific Northwest, the government owns mostly mountainous forests where nobody can farm, few could live, and lumbering is hard. However, these forests have never been cut and the ancient trees are huge. That makes them valuable.

The U.S. Forest Service allows private companies to lumber in the national forests. For years, lumber companies had plenty of ancient trees in their own private forests to cut. The old trees on public lands were safe from chain saws. However, by the late 1980s, lumbering companies had cut the last of their private ancient forests. They began to cut the public forests.

Conservation organizations saw that, in

10 years, the ancient forests would be gone forever. Loggers said they planted new trees as replacements, but conservationists argued that the new trees would never be a true forest. Instead, they would be a tree farm—crops to be harvested, replanted, then cut again. Meanwhile, the old forests and the species depending on them would be extinct. And so, conservationists sued the U.S. Forest Service over the northern spotted owl.

They argued that the endangered species depended on the ancient forests. Therefore, they said, the Forest Service broke the law when it allowed loggers to cut the forests.

For lumberjacks, the issue was frightening: some studies show that stopping the logging could eventually cost some 28,000 jobs. Loggers began displaying bumper stickers saying, "Loggers Are An Endangered Species, Too" and "Kill an Owl, Save a Logger."

Environmentalists argued that if logging

in the federal old-growth forests were allowed to continue, within a few years all the old trees would be cut and the loggers would be in the same fix they are now. But then the ancient forests would be gone forever. They also pointed out that many jobs were disappearing, not because of the owls, but because the logging companies were making their mills more efficient and needed fewer workers.

Judges agreed with the environmentalists. In 1990, the courts ordered a stop to the cutting of ancient forests on federal lands. Logging companies and their allies then began trying to open the federal ancient forests to logging once again.

For now, the northern spotted owl still flaps like a phantom in the shadows under the great trees. And, for now, the remnant of the continent's ancient forest still stands.

HABITAT CLOSE-UP

OLD GROWTH FOREST

When the Pilgrims landed at Plymouth, untouched forests stretched from the Atlantic coast to the Great Plains, and then beyond the plains to the Pacific. Now, in the lower 48 states, all but 5 percent of those primeval forests have been cut down. New forests have sometimes grown where the ancient forests stood, but they are different.

The ancient forests—old-growth forests—sprouted after the last Ice Age. Some California redwood trees are over 3,000 years old. More than 150 wildlife species depend for habitat on the ancient forests, with their towering trees and shaded ground underneath. But the huge trees are valuable to humans, too.

Just one ancient tree yields enough lumber for four houses. And the boards made from these tall trees are valuable because they have straight grains and few knots. As a result, loggers have been cutting the trees for centuries. Most of the remaining old-growth forest is in three states: California, Oregon, and Washington.

You can still walk in patches of old-growth forest. One is on the Olympic Peninsula in the state of Washington. Here you can sense the thousands of years this forest has stood.

Evergreens as broad as houses rise up around you, some 25 stories tall. Overhead, the branches and needles let in little light. Rain and fog blow in from the Pacific, and the moisture enables moss to grow everywhere. Here and there a shaft of sunlight slants down, heating the wet moss and sending up streamers of mist.

New trees take root in the moss atop a fallen tree's trunk. Years in the future, the fallen tree will be dissolved into the soil. But the seedlings that grew atop its trunk will be grown trees themselves. They stand in a row, showing where the fallen tree upon which they grew once lay.

20

Will Parrots Keep Talking?

Alex, an African gray parrot, has been teaching researchers at Northeastern University about parrot intelligence. Alex seems to have a lot of it.

"Tickle me!" Alex demands, when he is in the mood.

If he asks for a cracker and you give him a nut instead, he throws the nut on the floor: "I want a *cracker!*" he says. And when he wants to be sprayed with water from a bottle, he says: "Want a shower!"

Most parrots can mimic human voices, music, and other sounds. But the Northeastern University researchers believe that Alex *understands* a lot of what he says. For instance, Alex can correctly use words to identify colors, shapes, or materials, and he can count up to six. If the researchers show Alex a trayful of objects and ask him to pick the red

cardboard square, Alex will ignore the yellow cloth circle and the blue cardboard square and pick the red cardboard square. At least, he will if he feels like it. On days when he feels contrary he will pick every item on the tray *except* the red cardboard square. Or he may throw all the test objects onto the floor and tell his teachers: "Go away!"

Scientists who study parrots are amazed at their intelligence. That makes the plight of parrots particularly sad. Of the world's 332 species of parrots, 77 face immediate extinction. Most of the rest are threatened. One reason is habitat loss. Humans are cutting down the parrots' tropical forests for lumber and also to create farmlands. But parrots—with their appealing personalities and beautiful colors—also are disappearing for another reason: people crave them as pets.

The rarer parrot species become, the more collectors desire them. Certain species now sell for as much as $20,000 each. With so much money at stake, many people in poor countries are sneaking into the forests and illegally capturing parrots. Then they smuggle the birds into countries like the United States, where buyers will pay high prices for them. Such smuggling is destroying parrots.

GRAY PARROT

Each year, 8 million chicks are stolen from the wild. They are crammed into boxes, poorly fed, forced to live in their own droppings, then shipped in airliners to other countries. Four out of five of the baby parrots die. The few who do get through may be smuggled to individual buyers or sold in pet shops. The shop owners may be unaware that these parrots were illegally captured in the wild instead of being raised from eggs in captivity.

Meanwhile, parrot hunters know that parrots live in holes they find high up in trees. Such nesting trees are scarce, but parrot hunters often capture chicks by cutting down a nesting tree. Many chicks die when the tree falls, but a few may survive to be sold. Meanwhile, not only have the parrots lost their chicks, they also have lost a nesting tree. As poachers cut down ever more of the trees, it becomes difficult for parrots to reproduce.

One New York Zoological Society scientist has discovered another reason parrots are so vulnerable to bird collectors. He has spent hours hanging from tropical trees in a harness to observe South American macaws, a type of large parrot. He has discovered that they reproduce very slowly. His studies show that 100 pairs of macaws might produce as few as 20 young each year. When collectors steal their chicks, parrots cannot quickly replace them.

For the chicks that survive smuggling, life depends on their buyers. In the wild, parrots form close relationships with other parrots. Most parrot couples mate for life, which can be 70 or even 100 years. Observers have seen parrots fluttering over a dead mate, as if trying to revive it. Parrots spend months raising their chicks and teaching them how to find food and survive in the forest. Parrot families join with other parrots in large sociable flocks. They spend a lot of time screeching at each other among the treetops. Researchers say the screeching is often a way of communicating the location of good trees for nesting and gathering fruit.

Because of their need for close bonds, captive birds try to form close ties with their human owners. People who buy a parrot may be unprepared to give the bird the attention it craves. Also, they may feed it improperly, since even researchers are unsure about the right diet for every species. As a result of emotional and physical neglect, captive parrots often become lonely and depressed. Typically, such a captive lives only 5 years or so, instead of its normal 70 years.

Scientists concerned about the parrots' rapid disappearance are studying them wherever they are found—Latin America, Africa, tropical Asia, and Pacific Ocean islands. They also are experimenting with techniques for returning captive birds to the wild.

In 1986, for instance, the U.S. Fish and Wildlife Service confiscated 29 thick-billed parrots that smugglers had brought over the border from Mexico, where the birds are endangered. In the United States, the parrots once lived in Arizona, but they disappeared years ago. Fish and Wildlife officials decided to release the 29 confiscated parrots in the Arizona mountains, hoping to establish a new wild colony.

First, scientists had to glue new flight feathers on the birds' wings; the smugglers had pulled these feathers out so the birds could not fly to freedom. Officials hoped the artificial feathers would enable the parrots to outfly hawks, their greatest enemy. However, when they released the parrots, hawks quickly killed seven of them. Others flew away toward Mexico and vanished. The rest have been flying about in the Arizona mountains. They even built two nests and hatched eggs. But a ring-tailed cat (a relative of the raccoon) attacked one nest and, for unknown reasons, the parent birds abandoned the other nest.

Still, the tiny colony in the Arizona

mountains is hanging on. Wildlife Preservation Trust International and several government agencies are working together to bring back the thick-billed parrot. They are now raising thick-billed parrots in captivity and releasing them in Arizona.

Scientists are trying to save other species of parrots, too. If they can learn more about the various species and stop the illegal trade in captured parrots, they may yet save these exceptional birds from extinction in the wild.

21

How Long Will Piping Plovers Pipe?

It is sparrow-sized, its belly white and its back sandy colored, like the beaches where it lives. In spring, it wears jaunty black stripes on its shoulders and brow. You can miss the little bird because it blends with the sand and pebbles. But you might hear it.

The piping plover whistles a sad two notes. It sounds sad to us, at least. Perhaps that is because we know how few piping plovers are left and why they are dwindling.

Today these plovers, which have been described as tennis balls on spindly legs, are disappearing for the *second* time. Once they ranged up the length of the U.S. Atlantic coast and into Canada, on the Great Plains, and along the Great Lakes. But hunters nearly killed them off in the early 1900s. Protected by law, the birds recovered. However, since the 1940s, the piping plover has faced a dangerous competitor for its beach habitat: humans.

In March and April, since prehistoric times, the piping plovers have flown north from their winter homes in the Carolinas, Florida, and the Bahamas. Males stake out their nesting territories, running side by side down the beach to the water to set the "property" line. Then they try to attract the attention of females. They whistle, flutter in circles, scrape out sites for nests, and march in front of the females, stepping high. The females inspect the nesting sites the males are showing off and make their choice.

Piping plover nests are merely scrapes in the beach's sand and pebbles. They don't look like much, but that is the idea. The four pale eggs flecked with black blend so well with the nest's pebbles, sand, and bits of shell that they are invisible. Like camouflaged soldiers, the birds and their nests can vanish before your eyes. For eons, that was enough to save the species from predators. But it is a disadvantage with humans.

People are increasingly numerous, and we

PIPING PLOVER

like beaches. Unknowingly, as we run up and down the sand, we disturb the nesting plovers. We may not mean to bother the little birds, but we cannot see them.

We frighten the birds off their nests and their eggs cool. Sometimes we step on their eggs. Pet dogs and cats on the beach attack the birds. Trash left on the beach attracts gulls, raccoons, skunks, and other predators that eat piping plovers and their eggs. As a result, along the Atlantic coast, plovers are down to just 800 pairs.

Wildlife officials now close some beaches where the plovers nest. A few people become angry because they cannot use the beaches.

But many others understand why the beaches must be protected during the nesting months. For instance, at New York City's Breezy Point beach, about 40 piping plovers nest within sight of the Empire State Building. Since 1989, wildlife officials have closed the beach during the nesting season. As a result, the plovers are rearing five times more chicks.

Still, the plovers are barely holding their own. Around the beaches of the Great Lakes, only about 40 piping plovers remain and their future is in doubt. Meanwhile, the plovers that nest on the Great Plains have a different problem.

Here piping plovers once nested in large

numbers on sand bars in such rivers as the Platte. Spring floods scraped off plants that had grown on the sandbars, enabling plovers to nest on them again. But humans have been damming streams and rivers in the arid west, creating reservoirs of water for irrigating crops. As a result, the Platte River has shrunk to just a shadow of its original flow. Floods no longer clear vegetation from sandbars so the piping plovers can build their nests in the sand. When the plovers do find a place to nest, the shallower water makes it easier for predators to find the eggs. And sometimes dam managers release water from the reservoirs, which rushes down the river and washes away the plovers' nests.

You can help the piping plovers. At beaches, keep your dog or cat on a leash. And take away all your trash so that it will not attract predators. We may yet keep the piping plovers piping, singing their sad two notes.

HABITAT CLOSE-UP

SEASHORE

More goes on at a beach than we see.

A few plants, like sea oats, grow in the sand. Gulls and terns glide overhead. Shorebirds run to the water's edge to peck at tiny crabs. But other animals—invisible to us—live under the sand.

Each sand grain was once something else. Millions of years ago, it may have been an inland rock, weathered by rain, frost, and wind. Particles from the rock washed to the sea in rivers or blew on the wind. Waves threw the bits of rock onto the shore, creating a beach. On some beaches, sands may be pulverized shells or coral; sands can be fine or coarse, tan, white, or even black.

Certain worms build tunnels in the sand, gluing together the sand grains with secretions from their bodies. One species, the sandworm, is green with many segments and bristles. It grows to be nearly two feet long. At night, it leaves its tunnel to swim after small sea creatures, seizing them in its sharp jaws.

There is another hole in the sand; it belongs to the fiddler crab. Sometimes you see it skitter away sideways. Its cousin, the ghost crab, is colored like the sand and can disappear before your eyes. Where water washes over the sand, clams burrow. They stretch up a long tube to the surface so they can breathe and take in food. Starfish hide under the sand, waiting for their prey. Whelks—which resemble snails—move over the sand when the tide is out, hunting buried animals to eat.

Each beach has zones. High up, the sand is always dry. Farther down, waves wet the sand at high tide. Lower still, the sand is always moistened by the waves. And farther down, the sand is always submerged.

Each zone has its own species. Most of them we never see, for they are living in their secret cities under the sand.

22

Gone Forever

■ Wildlife ecologist Anne LaBastille stood beside Lake Atitlan in Guatemala, watching a ducklike bird swim by. It was black with a white ring around each eye. Its thick bill was white, ringed with black. The bird looked perky.

"It's a 'poc,' a funny duck that can't fly," a Mayan Indian told LaBastille, a visiting American.

That day in 1960 was LaBastille's first meeting with the giant pied-billed grebe. The Mayans named the flightless bird for its call: "poc." It lived only in Lake Atitlan.

LaBastille counted 200 giant grebes. Four years later she returned to the lake with its three volcanoes rising on the far shore. Only 80 of the grebes were left. Why had the population plummeted?

Now 32 vacation chalets had joined the original 12 Mayan villages along the lake's shore. But that did not explain the loss. The 15 miles of shoreline reeds—where the grebes

nested—were still intact. And Mayans rarely hunted the grebes.

Then LaBastille learned that local tourism officials and a U.S. airline had put 2,000 largemouth bass fingerlings in the lake. They hoped the fish would attract vacationers interested in sport fishing. One recently caught bass had weighed 25 pounds. Largemouth bass eat small fish and crabs—the grebes' food. The bass were depleting the grebes' food supply. Probably they also ate baby grebes.

LaBastille found a grebe's nest: a 100-pound mound of reeds. A fluffy chick, striped like a zebra, hopped into the water and climbed onto its waiting mother's back.

Saving the grebes became LaBastille's mission. She persuaded Guatemala's government to help. She walled off a shallow cove where grebes could nest safe from bass. She enlisted area residents as game wardens and became one herself. She held educational meetings to teach the Mayans how to protect the grebes. She even got the government to issue a giant-grebe postage stamp.

By 1968, the grebes had increased to 125 birds. LaBastille—by then known to the Mayans as "Mama Poc"—returned to Cornell University to work on her doctoral degree. When she heard that Guatemala planned a hydroelectric project at the lake that would lower the water level 23 feet and destroy the grebes' habitat, she fired off letters. Guatemala agreed to stop the project.

Meanwhile, the bass had grown too numerous. They had little left to eat but each other. The number of bass in the lake suddenly dropped. It was good news for the grebes: they increased their numbers to 232 birds.

Then, in 1976, an earthquake that killed 23,000 Guatemalans opened fissures in Lake Atitlan's bottom. The water level dropped until only two feet remained in the grebe's bass-free sanctuary. It was too little for the reeds the grebes needed for their nests. The wardens had to let the birds go into the lake.

Vacationers discovered the lake. Condominium towers and weekend homes sprouted along the shores. Reed beds were quickly vanishing. As a result, the grebe count dropped to 130 birds. A revolution erupted in Guatemala. One night, an unknown person shot to death Lake Atitlan's chief warden. By 1984, the lake's water level was lower than ever. Fishermen could now afford new nets that were entangling grebes. Vacation homes

PIED-BILLED GREBE

were increasing and so was pollution. Only 55 giant grebes remained. Ecologists know that when a species' population drops below 50 individuals, it is almost certainly doomed.

In June 1987, LaBastille received a letter from Guatemala. It informed her that the giant grebe was extinct.

Since then, she has written a book about her struggle to save the giant grebe: *Mama Poc: An Ecologist's Account of the Extinction of a Species.* She hoped that the loss of the giant grebe would at least be a warning that other species, too, could disappear. For the poc is only one of a long list of birds, beginning with the dodo, that humans have annihilated.

In America's early days, the continent's most numerous birds were the passenger pigeons. When they migrated, 100 million birds would fill the sky, covering the sun, their wings beating with a roar. Hunters killed millions. In 1914, Martha, the world's last passenger pigeon, died at the Cincinnati Zoo.

In 1875, a hunter shot a black-and-white bird—the world's last Labrador duck. In 1914, when the last passenger pigeon died, the Carolina parakeet also became extinct. In 1932, on the island of Martha's Vineyard off the Massachusetts coast, the heath hen disappeared forever. At least 32 Hawaiian bird species have disappeared.

Not long ago, a 10-mile stretch of Florida marshlands near the Cape Canaveral space center was home to 6,000 dusky seaside sparrows. These marshes were the species' only home. Here the small birds, dark on their backs, streaked on their bellies, ate insects and seeds. But humans grazed their cattle in the marshes. They built canals and roads, drained land and burned marsh plants to create fields for agriculture. The also sprayed the marshes with insecticides to control mosquitoes.

The last dusky seaside sparrow was a male named Orange Band. On June 16, 1987, at 12 years old, Orange Band died in a Florida zoo.

SECTION III

Something Fishy Is Happening to Fish

There are more than 20,000 *known* species of fish. Biologists estimate that the actual number of fish species may be 35,000. Fish are cold blooded and all extract oxygen from water using organs called *gills*.

Scientists divide fish into three classes. The oldest, *jawless fish*, such as hagfish and lampreys, evolved 500 million years ago. These primitive fish have no scales and no hard jaws.

Cartilaginous fish have skeletons made of cartilage rather than bone. Cartilage, which is softer than bone, is what stiffens our own noses. Fish in this class, such as sharks, arose about 400 million years ago. They have jaws, sharp teeth, and scales.

Bony fishes, with skeletons like ours, are newer members of the fish family. They number about 20,000 species. Unlike their primitive kin, these fish have an inflatable "swim bladder," like a small balloon, inside their bodies. By inflating their swim bladder, or letting air out, they can rise or sink in the water.

Worldwide, many fish species are waning. Sometimes it is because of too much fishing. Pollution also kills fish. So do dams that change water levels or prevent fish from swimming up rivers to mate.

Sometimes fish invaders kill other fish. Africa's Lake Victoria, as big as Switzerland, once was home to 300 species of fish. Then, in the 1950s, Nile perch got into the lake; nobody knows how. These big perch have decimated the lake's native fish. About 200 of the 300 original species are now extinct. As a result, the lake's ecology—the interactions of the species—is disrupted.

When the Nile perch eat up all the native fish, they will have nothing left to eat but their own young. Meanwhile, microorganisms are multiplying in the waters because the small fish that once ate these tiny animals and plants are gone. The huge population of microorganisms is consuming the lake's oxygen. Eventually, the Nile perch will die off, too. Lake Victoria may one day be dead, with no living creatures at all.

But biologists are trying to save the lake. They are breeding the lake's remaining native fish at aquariums and laboratories around the world. Also, a new commercial fishing industry for Nile perch may eventually help control the invaders. Then scientists can release back into the lake the native fish they have raised in captivity. If all goes well, at least some of Lake Victoria's native species may swim once again in their natural habitat.

23

Pacific Salmon Are Swimming Uphill

Pacific salmon can swim hundreds of miles against strong currents. They can even leap up waterfalls. But now salmon face hurdles they cannot overcome.

All six species of Pacific salmon are disappearing. Two species—chinook and sockeye—are now listed as threatened or endangered. To understand why, it is necessary to know the salmon's amazing life story.

Consider the chinook.

In Idaho, in the spring, reddish salmon eggs lie strewn across the gravel bottom of a pool in a trickling creek. They hatch into tiny fish. The little chinook are 800 miles from the Pacific. But when they have grown to about six inches, they feel an urge: swim!

The little fish, or *smolts*, swim down their creek into a larger stream. From there, they follow the current into the big Salmon River, then into the even wider Columbia River. As the smolts near the ocean, the fresh water turns salty. But their little bodies have been changing, too. By the time they swim out into the Pacific, they have become ocean fish, adapted to salt water.

For about three years, the chinook swim over 2,000 miles of the Pacific. They feed on smaller fish, growing ever larger—up to 100 pounds. They look like silver torpedoes. Then one autumn, an inner urge makes each chinook swim eastward, toward Oregon.

The chinook swim into the Columbia River and travel hundreds of miles upstream to the Salmon River. They swim desperately against the heavy current, jumping through rapids and leaping up waterfalls. Finally, they come to the tiny creek where they began life. Somehow, the chinook recognize their birthplaces. Perhaps it is the smell of the water. The fish swim to the spot where they hatched.

By now the chinook look weary and tattered. The males have turned dark red, their jaws withered. The fish are dying. But they must do one more thing: the females lay

PACIFIC SALMON

eggs in the gravel and the males fertilize the eggs by covering them with sperm. Then the chinook gasp wearily in the shallows until they die. Their decomposing bodies provide nutrients for their eggs.

In the spring the eggs hatch and tiny salmon emerge. The cycle begins again.

That is how it always has been. But no more.

Once, 10 million salmon swam up the rivers of the Northwest every year. Now, fewer than 2.5 million salmon make the trip. Of those, only 500,000 hatched in the wild. The rest are hand raised in tanks and released for fishermen to catch. But these tame salmon cannot withstand diseases and dangers as well as wild salmon. One biologist calls them "fish

wimps." Meanwhile, wild chinook and sockeye salmon are disappearing fast.

About a third of the salmons' streams are now blocked by dams or degraded by pollution and logging. Salmon only breed in their native stream: if a dam blocks it, that stream's salmon—called its "run"—become extinct. So far, about half of the Pacific salmon runs are gone. Now northwesterners are battling over what to do about the salmon.

Salmon supporters believe sacrifices are necessary to keep the salmon from disappearing. As Idaho governor Cecil D. Andrus, a friend of the salmon, has said: "We ought to be ashamed of ourselves if we can't save them."

Everywhere in the United States, rivers and streams are ailing. Often, the "illness" is pollution.

Water is powerful. Over centuries, a river can dig out ravines and vast valleys. Yet, a great river may begin as a mere trickle. Trickles join to form streams. Streams combine to create rivers.

Algae and mosses grow on stones in the stream. All sorts of insects live among the water plants. So do other species, like crayfish and fish. Grasses and trees growing on the river's moist banks may be different from plants growing nearby. In fact, plants on a river's cooler and darker north-facing bank may be different species than those that grow on the opposite, sunny bank. Many creatures seek out rivers and streams: birds, frogs, snakes, salamanders, otters, mink, muskrats, beavers, raccoons.

Humans are drawn to rivers, too. Rivers provide water for drinking and for crops. They are highways for boats. Cities may discharge their wastes into rivers. The moving water can power mills and factories and generate electric-ity. But humans often pollute the rivers they use.

For instance, chemicals called PCBs were once used in electrical equipment. Factories flushed PCBs into rivers, where they poisoned wildlife. In 1977, the government banned PCBs. But it was too late: the PCBs were imbedded in the mud and sands of river bottoms. In some places, fish are now so full of PCBs they are unsafe to eat.

Sewage is another type of pollution; so are farm pesticides that get into the water. And factories may release other chemicals into streams and rivers. Logging can erode stream banks.

Another problem is dams, which can slow a river's flow and block migrating fish. Right now, dams are contributing to the decline of 68 percent of the species that have lived for millions of years in the rivers of the U.S. Southwest.

Conservationists are now trying to preserve our rivers and streams. If they succeed, we will not see the last of fish like the smoky madtom, the blackside dace, and the Pahrump killifish.

24

The Razorback Sucker Is on the Edge

No fish looks stranger than the razorback sucker: behind its head, its body bulges up-ward. It *looks* as if one committee designed its head, another its body.

Actually, the razorback sucker's high-rise back stabilizes it in gushing western rivers. For eons, the two-foot-long fish thrived from Wyoming to Mexico, seeking spots where warm water flows over sand, gravel, or rocky bottoms. It fed on tiny plants and insects. Razorbacks were once so common that a single fisherman in 1949 caught six tons of them in one lake along the Salt River. But the razorback no longer thrives.

Dams have decimated the species. Since 1910, 15 dams have gone up on the Colorado

RAZORBACK SUCKER

River and its tributaries. The dams changed the flow and temperature of the water, making it harder for the suckers to survive. Meanwhile, the dams enabled alien species to move in—carp, channel catfish, red shiners, largemouth bass, walleye, and northern pike. These invaders eat the razorback sucker's eggs and young and compete with it for food. As a result, the razorback population is now down to just a third of its original range.

Suckers can live 30 years or more. Biologists believe many of the razorbacks now alive are old-timers, slowly dying off.

The U.S. Fish and Wildlife Service is trying to restock the endangered fish in suitable rivers. So far, however, the biologists see little sign of success. They are still trying, hoping the fish that seems designed by an inattentive committee keeps on swimming.

25

Extinctions Long Ago

On the plains of Nebraska, huge camels extend their giraffelike necks to graze on tree leaves. Rhinoceros herds wade shoulder deep in a stream.

It is Nebraska 10 million years ago. Rhinoceroses were then North America's most common large plant eaters. Giraffe-camels wandered the plains. Now both species are gone. So are the elephants—woolly mammoths and mastodons—that once trumpeted in American forests. Saber-toothed cats are gone, along with giant sloths and beavers as big as ponies. All are extinct.

In Argentina, scientists have found the remains of a bird with the wingspan of a jet fighter. Eight million years ago, South American forests were home to a crocodile the length of a tractor-trailer truck. These, too, disappeared. Scientists are unsure what caused all of these species to die, but they know it was not humans. They do have some ideas.

For one thing, by studying rocks and fos-sils, scientists have learned that every 26 to 30 million years great extinctions occur. In each of these episodes, 50 to 95 percent of all the species living on Earth vanished in a relatively

MAIASAUR

short time. Then evolution begins again. New species emerge and multiply, until yet another of the mysterious mass extinctions wipes out large numbers of them. Research now suggests that these periodic mass extinctions may be caused by a variety of catastrophes.

For instance, 250 million years ago—before the dinosaurs came to dominate the planet—something killed up to 95 percent of all species then alive. Ocean creatures died in vast numbers and so did land animals. Recently, scientists discovered that during that time volcanoes erupted in Siberia, an eruption that went on for 200,000 years. It was so massive that the lava flowed over 130,000 square miles. Many researchers now suspect that this huge volcanic explosion triggered the extinctions.

A volcanic eruption so huge would have spewed millions of tons of dust, ash, and gases into the atmosphere. That would change the weather. When the climate changes—becoming cooler or warmer, drier or wetter—the delicate web of interactions among plants, plant-eating animals, and predators is disrupted. Many species cannot adjust to the loss of habitat and food. They die off.

And what about the dinosaurs? We humans have been around only about a million years, but dinosaurs lasted 160 million years. They dominated the seas and the land. Then, about 65 million years ago, the dinosaurs died off. Scientists are trying to figure out why.

They have many theories. For instance, some researchers believe an epidemic swept the prehistoric world, killing dinosaurs by the millions. Others believe vast volcanic eruptions changed the climate. But, so far, the evidence seems to favor an even stranger theory: that a huge space object—such as a comet or asteroid—slammed into the Earth.

Scientists have several clues. For instance, the layer of rocks from the period when the dinosaurs disappeared contains large amounts of the metal iridium. Rare on Earth, iridium is common in asteroids and meteors. That suggests that a huge meteor or asteroid struck the Earth and exploded with the force of many hydrogen bombs, raining its iridium across the planet. The explosion doomed the dinosaurs.

Some scientists argue that the impact would have billowed dust into the atmosphere, darkening and cooling the planet. Until the air cleared, plants stopped growing. Dinosaurs died from chilling and lack of food. Other scientists believe the impact would have set off intense firestorms worldwide: the dinosaurs broiled alive.

Recently, planetary scientists discovered *two* huge craters from a prehistoric impact, one under the Caribbean and the other on Mexico's Yucatan Peninsula. They believe the impact occurred just when the dinosaurs were disappearing.

Meanwhile, paleontologists (scientists who study fossils) have found evidence in North Dakota and Montana that the dinosaurs disappeared relatively quickly. They could not pinpoint the duration of the extinctions: it might have taken 100,000 years or just a few weeks. But their finding further supports the impact theory. Diseases would presumably have required millions of years to eradicate the entire dinosaur population.

Prehistory's extinctions are unexplained; many scientists still dispute the impact theory. But the reason for *today's* extinctions are clear: we know the culprit and it is us.

Amphibians and Reptiles Are Earth's Old-Timers

When the Earth was younger—350 million years ago—the most advanced organisms on land were plants and insects. In the sea, it was fish. But a new creature evolved in the ocean and crawled onto land. It was an animal partly of the water, partly of the land: an *amphibian*. Since then, amphibians have changed little. And they are still with us, including toads, frogs, salamanders, and newts.

Amphibians begin life underwater, hatching from eggs into an immature larval form, such as a tadpole. Tadpoles and other amphibian larvae extract oxygen from the water with gills, like fish. As the amphibian matures, it develops lungs. Adult amphibians breathe air through their lungs, as we do.

About 2,500 species of amphibian are alive today. Some are no larger than your pinkie's fingernail. But a foot-long frog in west Africa weighs as much as a house cat. A Japanese salamander measures five feet from snout to tail. Some tiny South American frogs, colored like gems, ooze deadly poisons.

About 300 million years ago, some long-vanished amphibian began to change. Instead of laying its eggs in the water, it deposited them on land. Its hatchlings had lungs and could breathe air even before they were fully grown. It had a more efficient heart than its amphibian ancestors. And while an amphibian's skin is smooth and soft, it had hard, protective scales, like a Medieval knight's chain mail. This new creature was a reptile.

On the early Earth, the reptile was evolution's most up-to-date model, equipped to thrive in the tropical warmth. It quickly evolved into all sorts of species. Some of the new reptiles were tiny, scurrying in the underbrush. Others were enormous, the size of modern houses, nibbling the tops of evergreen trees. Some learned to swim and returned to the seas from which their ancestors, the amphibians, had originally come. Others developed wings and flew. For millions of years, these reptiles—the dinosaurs—dominated the land, oceans, and skies.

Eventually, the great reptiles died off. Scientists are not yet sure why. Perhaps they had difficulty adjusting to climate changes because they were *cold-blooded*. That means they relied on the sun for warmth. Modern amphibians are cold blooded. So are today's reptiles—snakes, lizards, turtles, and tortoises. Birds and mammals, which came to dominate the Earth after the great reptiles died off, are *warm-blooded*. By turning food into energy, they can keep themselves warmer than the surrounding air. As the climate cooled, the new warm-blooded mammals and birds had a better chance of survival.

Reptiles no longer rule the Earth. But when we see a painted turtle in a pond or a garter snake in the yard, we are—in a way—looking at our own evolutionary greatgrandparent. And many reptiles and amphibians are now endangered.

Sea Turtles Are a Vanishing Mystery

On a moonlit beach in Costa Rica, a green sea turtle rides the breakers in from the Caribbean.

She climbs ponderously up the beach, 300 pounds of shell and leathery skin. In the wet sand behind her, she leaves flipper prints. When she is beyond the reach of the waves, she digs into the sand.

All four flippers flail as she scrapes out a two-foot-deep pit, the size of her body. She sighs as she works, as if longing to return to the sea. At the back of the pit, she pushes with her rear flippers to dig a special chamber for her eggs.

It is a ritual eons old, this emergence of the green turtle from the sea. Once millions came on such moonlit summer nights. Now, only hundreds.

She has been skittish. A rustling on the sand or a stirring in the moonlight may send her back to the sea. Jaguars, South America's big cats, attack sea turtles. So do human hunters. But tonight the turtle sees no danger. She begins laying her eggs.

Now, nothing deters her, as if she were in a trance. As she lays her leathery eggs, each the size of a ping-pong ball, she will not notice if raccoonlike coatis steal them from beneath her. Wild dogs and pigs may seize the eggs. Even if a jaguar attacks the turtle herself, she will continue laying her eggs.

Worst among her enemies are human poachers, who savor turtle eggs and turtle meat. But she has guardian angels. Watching in the dark is a biologist. Also watching are tourists who have volunteered to help him study these endangered turtles. Federal guards are watching, too. For this 22-mile strip of beach is the "Tortuguero," the place of the turtles. It is the green turtles' major nesting site in the Caribbean, and Costa Rica has made it a national park.

Sea turtles need protection: although huge, they harm nothing. They never bite or scratch. If attacked, the worst they might do is desperately kick sand at their assailant.

When the turtle has finished laying her

SEA TURTLE

NEST

100 or so eggs and covering them with sand, the biologist clips a tag to one of her flippers. Biologists know little of sea turtle life. If anyone finds her in the coming years, the tag will give researchers clues to this turtle's movements. Once she is tagged, the biologist measures her. Then she plods back down the beach. Waves wash over her. Now she is afloat, swimming out through the breakers to resume her mysterious life in the sea.

Before Europeans settled the New World, 50 million green sea turtles swam in the Caribbean. Now, only about 20 thousand greens nest at Toruguero. However, scientists who study sea turtles hope to use the information they glean to help the turtles survive.

For instance, scientists now know that after sea turtle hatchlings dig their way out of the sand they immediately race to the sea. They spend the rest of their lives in the ocean, feeding on plants in the calm shallows off coasts. They may swim thousands of miles from their hatching place. Sea turtles can swim up to 20 miles in one day. One tagged turtle swam 525 miles in a month.

Males never leave the ocean. But females come ashore once each year, on a summer night, to lay their eggs. Scientists believe they always return to the beach where they hatched, 10 to 50 years earlier. But how do they find it?

Some researchers suspect, although they do not yet know for sure, that turtles "smell" their way home. They theorize that a newly hatched turtle sniffs the particular chemicals of its native beach. Years later, when she is ready to lay her first eggs, the adult female follows tiny traces of those chemicals in the sea back to "her" beach.

Ever since dinosaurs roamed the continents, sea turtles have made these annual migrations to lay eggs. Now it is up to humans to help the turtles to survive. Biologists have been trying to save these ancient species in a number of ways.

For instance, at Tortuguero, in Costa Rica, when high water or predators seem likely to destroy incubating eggs, biologists dig up the eggs and hatch them in captivity. Then they immediately let the tiny sea turtles run down the beach and make their way out to sea through the breakers.

Green turtles are not the only endangered sea turtles. Other species in trouble are Kemp's ridleys, loggerheads, hawksbills, Indian sawbacks, Indian softshells, olive Pacific ridleys, and leatherbacks. All these species are declining for similar reasons. One is the international trade in sea-turtle leather and shells. Another is local people eating turtles and their eggs. Meanwhile, development and pollution destroy the turtles' nesting beaches.

Also, U.S. shrimp boats accidentally net some 48,000 sea turtles each year. The turtles are legally protected and conscientious fishermen try to release them. Even so, about one in four of the turtles caught in nets dies. Federal researchers have been trying to persuade shrimpers to use devices on their nets that keep turtles and other large creatures out. Many shrimpers resist, arguing that the "turtle-excluding devices" also allow large numbers of shrimp to escape from the nets.

Another danger to sea turtles is beach lights. When baby turtles hatch, they must scurry to the sea as quickly as possible to escape predators, such as raccoons and sea birds. Lights on a beach disorient them, so they cannot find the water. Now many seaside communities are dimming their lights during the nesting season out of concern for the little turtles.

Meanwhile, researchers have been trying to establish a new colony of the highly endangered Kemp's ridley turtles, taking eggs from their one remaining nesting beach in Mexico and transporting them to Padre Island, Texas. The researchers hope the little turtles will adopt Padre Island as their new home and return there to lay their eggs. That way, the world will have two hatching sites for this species, not just one.

Perhaps the future is not entirely grim for these remnants of the great Age of Reptiles. Where communities are protecting the turtles' nesting beaches, populations of at least some species seem to be increasing—a good sign.

27

Desert Tortoises: Helping These Living Bulldozers

Among the cactus and sand dunes of the American Southwest, the desert tortoise is slowly crawling. But is it crawling toward survival or extinction?

Some humans have endangered this ancient species. Now, other humans are trying to save it.

Tortoises are land-dwelling turtles. Instead of a turtle's webbed feet for swimming, tortoises have big feet like shovels, for digging. The desert tortoise is about as big as a football. It wanders the dry country seeking food, mostly plants and insects. Like all reptiles, the tortoise is cold-blooded and cannot survive extreme cold or heat. Yet, depending on the season and time of day, its desert habitat can be searing or frigid. To escape both, the tortoise digs deep burrows into the earth, using its lower shell's front edge as a bulldozer blade.

All sorts of threats have endangered the desert tortoise—livestock grazing, off-road recreational vehicles that destroy the desert habitat, spreading suburbs. Some people capture the slow moving reptiles for pets. Some kill them for "fun."

Now that tortoises are few in number, they are vulnerable to other threats that might drive them to extinction. In California's Mojave Desert, for instance, the growing human population has created new garbage dumps, which has led to a population explosion among ravens. Some of these large cousins of the crow have been attacking desert tortoises. To help the tortoises, government officials and the Humane Society of the United States have been keeping ravens away from their usual resting areas in the tortoise preserve.

But the tortoises face a worse enemy than a few rogue ravens. Thousands of desert tortoises have died from a mysterious breathing disease. Veterinary researchers at the University of Florida are now studying the

DESERT TORTOISE

disease, hoping to find a way to protect the desert tortoises.

Humans are helping tortoises in other ways, too. In Las Vegas Valley, Nevada, where housing construction has been destroying tortoise habitat, developers have joined with environmentalists and government officials to create a Desert Tortoise Conservation Center. The Center will become a safe home for tortoises displaced by human construction.

Desert tortoises face many threats. But, with help, the species still has a chance to survive.

28

Protecting the Dragons of Indonesia

In 1912, on Komodo and five other small, uninhabited islands of Indonesia, explorers discovered an astonishing beast: a fierce, greenish lizard as long as a car. They called it the Komodo dragon.

Up to 10 feet long and weighing 280 pounds or more, it is the world's largest lizard. Dragons feed on dead animals they find in their hilly islands' rain forests and grasslands. They also hunt monkeys, wild pigs, and deer.

At night, when the air is cool, the cold-blooded dragon sleeps. As the morning sun warms the island, the dragon rouses itself and lumbers into the forest to hunt. Flicking its forked white tongue in the air, it senses chemicals emitted by its prey. The dragon crouches behind bushes or a rock, waiting. When its prey comes near, it charges, grabbing its victim with powerful jaws. Dragons have occasionally killed humans in this way.

Escaping a Komodo dragon is not easy. They can climb trees and swim. In fact, the Komodo dragon is its island's biggest, strongest species. It has only one real natural enemy: an even bigger Komodo dragon. Sometimes the dragons fight over dead animals. And sometimes big dragons eat smaller dragons.

Dragons mate in early summer. A month or so later, the female digs a burrow up to 30 feet deep. Inside, she lays a dozen or so leathery eggs and goes on her way.

Newly hatched dragons are eaten by birds, snakes, and even adult Komodo dragons. For safety, they spend their early weeks hiding in trees. However, they grow about a foot a year. By the time they are six years old and six feet long, their only threat is from older and larger dragons.

Yet, Komodo dragons are endangered. That is because they were taken by collectors of exotic animals. Now they are protected by law.

For now, at least, the Komodo dragon is

KOMODO DRAGON

holding its own. That means that on the uninhabited small islands of Indonesia, it still is possible for visitors to think they have stepped back into the age of dinosaurs, when huge reptiles ruled the world.

29

The American Alligator Is Back

Night in a Florida swamp.

Two biologists glide a boat across the black water. Shadowy oaks festooned with Spanish moss lean overhead. An owl hoots. Paddles splash. One biologist turns on a flashlight.

Red eyes.

All around the boat, red eyes watch from the water. The swamp is alive with alligators.

That is how it is today in many southern swamps. Yet, just 20 years ago, the alligator was slipping fast toward extinction. Hunters and trappers had taken so many of the big reptiles for meat and skins that alligators seemed doomed. But in 1967 the alligator was added to the federal endangered species list. That was its salvation. Hunting stopped, and the alligator began to return to the South's swamps and waterways.

Florida's dusky sparrow became extinct because it could not adapt to changes in its habitat. Florida panthers have lost so much of their range that their future is doubtful. But the alligators' story is different because they are so adaptable. As developers drained swamps to build houses, alligators began to turn up in suburban swimming pools, storm drains, and supermarket parking lots.

Once protected from human guns, alligators relied on their own survival tricks. For instance, where water levels drop, alligators dig deep underground chambers that fill with water. There the cold-blooded creatures can avoid extreme heat or cold. Young alligator hatchlings, smaller than a human's foot, feed on frogs, tadpoles, insects, and other small prey. Adult alligators grow to be as long as a rowboat by the time they are about six years old. They eat almost anything they can catch: raccoons that come to the shore to drink, swimming muskrats, snakes, or birds. Alligators sneak up on their prey and drag it down into the water to drown.

The female alligator builds a nest from rotting twigs, leaves, branches, and mud. She lays her eggs in the nest and then covers them with the vegetation. As the nest material rots,

AMERICAN ALLIGATOR

it produces heat that incubates the eggs. Many animals and snakes would like to eat the alligator's eggs. However, one reason the species could spring back from near extinction is that mother alligators guard their nests during the three months or so the eggs are incubating. If any creature tries to snatch the eggs, the mothers ferociously fight it off.

When they are ready to hatch, the baby alligators screech, which tells their mother it is time to open the nest. The hatchlings immediately crawl into the water and go off on their own. Every year they will grow a foot. But when they are small, alligators are often eaten by predators. If they survive to adulthood, they may live as long as 50 years and grow up to 18 feet long. Such huge reptiles have no enemies except humans. And, given human protection, they have now come back. In 1987, the government removed alligators from the threatened and endangered species list.

But the alligators' return has caused problems. Over the past 20 years, Florida's human population has doubled and so has the state's alligator population, which is now over one million. Today, humans are building houses on land that just recently was alligator habitat.

As a result, alligators sometimes appear in people's yards and attack their pets. About a dozen times a year, alligators attack humans. And so Florida officials now remove troublesome alligators and allow limited hunting to keep the number of alligators under control. Meanwhile, they are teaching humans how to live with their alligator neighbors. With understanding from humans, they believe, the two species will be able to live together.

30

What Is Happening to Amphibians?

■ Frogs, toads, and salamanders—the amphibians—swam in the Earth's waters and walked the land long before dinosaurs appeared. Dinosaurs are gone, but the amphibians have held their own. Yet, in just the past few years, scientists have discovered that amphibians are mysteriously disappearing.

It is happening around the world:

- In 1987, in one month of study at the Monteverde Cloud Forest reserve in Costa Rica, scientists found 1,000 brilliantly colored golden toads. In 1990, they could not find a single golden toad.

- Australia's gastric-brooding frog was so abundant just a few years ago that biologists could catch 100 in one night of collecting. Today, the entire species is gone.

- California's High Sierra mountains swarmed with yellow-legged frogs when biologists studied 38 ponds there in the 1970s. In 1989, biologists went back to recount the frogs. The creatures had vanished from all but one of the ponds.

- In the U.S. Northeast, about two-thirds of the amphibian species are now either uncommon, threatened, or declining.

- Two species of toad—the American and Fowler's toads—have now disappeared from around Pittsburgh, Pennsylvania, where they once were plentiful.

Many other species, from the Chinese giant salamander to the Wyoming toad, are now endangered. Scientists are particularly concerned because the amphibians are not just disappearing from urban and suburban areas. They also are vanishing from preserves and isolated sites where few people go. Could the disappearances be a warning that something is seriously wrong with the natural world?

Amphibians are good indicators of environmental disturbances because of their life cycle. They spend part of their life in water and part on land. Also, their eggs are sensitive to chemicals in water. As adults, they are still vulnerable to pollution because they absorb and release gases through their skin.

Biologists are so worried about the vanishing amphibians that in 1990 they held a scientific summit meeting on the disappearances. Since then, they have set up a task force to find out why amphibians are disappearing. They suspect there may be more than one reason for the declines.

Water pollution from industry is probably a major cause. Another is *acid rain*; smokestacks from factories and power plants spew chemicals into the air that turn rain acid. The falling acid rain makes streams, ponds, and lakes more acid, which kills amphibian hatchlings. Other amphibians disappear because the spreading human population takes away their habitat. One study has shown that logging in the Pacific Northwest often obliterates salamander populations for up to 40 years after-

COMMON TOAD

wards. When forests are cut down, hillsides erode and pour silt into salamanders' streams.

Some amphibians have declined because states put trout into lakes for fishermen to catch. The trout eat up amphibian eggs and tadpoles. Many Asian frog species have declined because they are hunted for food.

"If they're dying off, there is a message there for us," according to biologist David Wake, of the University of California at Berkeley. "Amphibians are tough critters. They survived whatever wiped out the dinosaurs, and they have thrived through the age of the mammals. If they start to check out now, we'd better take it seriously."

Invertebrates Are Creatures with No Backbone

Uertebrates are animals with backbones, or spines. We humans are vertebrates. So are the other "higher" animals, from sunfish and garter snakes to bald eagles, bottle-nosed dolphins, and white-footed mice. But we vertebrates are outnumbered by *invertebrates* —creatures with no backbones.

Most animals are invertebrates. Honey bees, black widow spiders, lobsters, snails, clams, worms, jellyfish, octopuses, and squid are all invertebrates. They have no backbones. Yet, they manage to hold themselves together.

Some invertebrates, like jellyfish, have no skeletons at all. They float in the water like living bubbles. Others, like insects, clams, and snails, wear their skeletons on the *outside* of their bodies. Their skeletons take the form of an armorlike shell, which supports the animal's body and protects it.

Fewer invertebrates are on the government's endangered and threatened species lists than vertebrates. That is not because few invertebrates are endangered, however. It is partly because we know less about invertebrates. In fact, scientists believe they have not yet even discovered millions of species of invertebrates. Also, invertebrates appeal to us less. They are not furry like pandas or beautiful the way we think eagles are beautiful. Many are small and easy to ignore. But we put ourselves in danger if we ignore endangered invertebrates.

Animals, plants, and microbes interact in ways that are so complex that scientists are only beginning to understand them. For instance, certain ocean plants, almost too small to see, produce much of the atmosphere's oxygen. If those plants ever become extinct, most life on Earth would soon end. Invertebrates are vital, too. Without insects and other invertebrates to eat, for example, most birds would starve. Tiny ocean invertebrates, called *krill*, are the food of some of the huge whales. Many of these invertebrates are disappearing at a frightening rate.

Consider insects. Scientists know of more insect species—about 751,000—than any other animal or plant. But they estimate that up to *10 million* insect species are still undiscovered, many living in tropical rain forests. Those tropical forests are now rapidly disappearing, logged for lumber or burned to create pastures and farmlands. As a result, scientists at Cornell University estimate that about half of all insect species may become extinct over the next 30 years. Many will become extinct before we even discover them. In fact, scientists estimate that we are now losing 19 insect species *every hour*.

31

If It Could Sing, the El Segundo Blue Would Sing the Blues

Where Los Angeles, California, now sprawls, sand dunes once rolled down to the Pacific. Yellow buckwheat grew on the dunes. And here lived the El Segundo blue butterfly, just an inch across, fluttering on lavender-blue wings.

Caterpillars of the El Segundo blue ate only buckwheat. And as Los Angeles grew across the dunes, the buckwheat disappeared under freeways and housing subdivisions.

Today, the city has covered all the dunes except two tiny patches—80 acres at the end of a Los Angeles airport runway and two acres at a Chevron company oil refinery. Buckwheat still grows on these small plots. Here the El Segundo blue butterfly hangs on, maintaining its ancient life cycles.

Adult El Segundos live only seven days. After mating, the female selects a flower and "dances" on the blossom to make sure it is suitable. Then she lays her eggs among the petals. About a week later, the larvae hatch. Each little caterpillar is pearly white, the color of the fresh buckwheat flowers it eats. That

EL SEGUNDO BLUE

way, the caterpillar can hide from birds and other predators that might eat it. As the caterpillar eats buckwheat and grows over the next month, it goes through five stages. It changes color to match the buckwheat blossom's seasonal color changes, from white in August to creamy yellow or pale reddish-brown in September.

When it is ready, the caterpillar crawls to the sand at the base of the buckwheat plant and enters the mummylike pupa stage. Slowly, the caterpillar changes into a blue-winged butterfly.

Humans do what they can to keep the species alive. Chevron has fenced in the dunes at its refinery to keep out off-road vehicles and butterfly collectors who might harm the El Segundos. The airport has fenced in its patch of dunes, too. At both sites, biologists cultivate buckwheat. Without buckwheat to eat, the El Segundo blue cannot live.

Even with help, the El Segundo is struggling on its slivers of remaining habitat. The butterflies are so few that any small incident could trigger extinction—a small change in the climate, perhaps, or a predator insect accidently brought in from overseas on the landing gear of an airliner. For now, though, the El Segundo still flutters among the buckwheat, on lavender-blue wings.

32

The Gravedigger Bug Digs In

Andrea Kozol looked like a tourist boarding the ferry on the New England coast. She carried camping gear and a picnic cooler. But her cooler contained no sandwiches—inside were 50 American burying beetles, among the last of their kind.

Kozol, a young biologist, had raised the endangered beetles in her laboratory at Boston University. Now she was taking them to an island, hoping to give the insects a new start in the wild.

American burying beetles are large by beetle standards—about the length of a paper clip. They are jet black, splotched with orange. Once they thrived throughout the eastern half of the United States and parts of Canada. But about 30 years ago they began to disappear. Now scientists can find them only in a few spots in Oklahoma and on Block

Island, off the Rhode Island coast. Exactly why the burying beetles have disappeared so rapidly puzzles researchers like Andrea Kozol. They suspect it must have something to do with the beetles' peculiarities.

For one thing, unlike most insects, mother and father burying beetles are good parents, protecting their young and feeding them. Also, they have odd dining habits.

Burying beetles search at night for small dead animals—mice or pheasant chicks, for instance, or a dead garter snake. When the chemical sensors in the beetles' antennae detect a carcass's smell, they converge at the site. Male beetles fight with other male beetles over the carcass. Meanwhile, the females also fight among themselves. The largest beetles drive away their opponents so that only the biggest male and the biggest female are left. They scout the terrain around the carcass for a good burial site, and then they go to work.

The two beetles crawl under the carcass, upside down. Lying on their backs, their feet on the carcass, they work their legs to "walk" the carcass forward to its grave site. The carcass might be 200 times the beetles' own weight, which is like a human couple moving a tractor trailer truck. The beetles bury the

BURYING BEETLE

off mouthfuls from the carcass, digest the meat, and then regurgitate it to feed to their young.

Every day, cared for by their doting parents, the larvae grow 10 times larger. Meanwhile, their parents feed them and fight off any other insects that may invade their underground home. When the growing young are about four days old, their father leaves and goes on his way. Their mother continues to care for them for about one more week. By then, the carcass is just bones. Now the mother leaves her well-fed larva on their own. They burrow into the ground for a month to "pupate"; looking like insect mummies, the grublike larvae change into adult beetles.

Then the new American burying beetles go off on their own, hunting for dead creatures to eat. By winter, they are ready to hide away and sleep through the cold months. They re-emerge in the spring to begin the cycle again.

That is how it always has been. But in the last few decades, something unknown has wiped out most of the American burying beetles. Scientists wonder, was it a change in the climate? Or was it pesticides sprayed on farms and lawns? It seems unlikely, because other species of burying beetle still thrive. Biologists also question whether it was loss of habitat that eliminated the beetles, because the loose soil they need to bury carcasses does not seem to have disappeared. Perhaps development has eliminated so many small animals that the beetles no longer find enough car-

carcass about four inches underground. Then they get ready for family life.

First, they dig out a nursery chamber for their young next to the carcass. Then they remove the carcass's fur or feathers and roll it into a ball. They cover it with preservative secretions from their bodies.

"When they're done, the carcass looks like a giant meatball!" says Andrea Kozol.

After the beetles are done preparing the carcass in their underground house, the female lays about three dozen eggs in one chamber. The wormlike larvae hatch less then a week later and crawl to the top of the carcass, where they cluster like little birds in a nest, waiting to be fed. Much like birds, their parents bite casses. The scientists say that is just a theory; they do not yet know why the American burying beetle is almost extinct. But they are trying to bring the species back.

That is why, a few years ago, Andrea Kozol worked out a plan with federal and state conservation authorities. She removed 20 of the approximately 500 beetles still living on Block Island to her Boston University laboratory. She was delighted to find that it was easy to breed the American burying beetles in captivity. When she had 50, she took the beetles in her picnic cooler to Penikese Island, a wildlife sanctuary off the New England coast.

She released the 50 beetles and provided a chick carcass for each pair. Then she left, hoping the beetles raised in captivity would survive in the wild.

Ten days later, Kozol went back to Penikese Island. She dug up 17 of the carcasses: more than half contained larvae!

Until the biologists can determine for sure why the American burying beetle is disappearing so rapidly, the species will be on the brink of extinction. But Andrea Kozol's efforts to start new colonies mean these unusual insects may have a future.

33

If Mussels Had More Muscle, There Might Be More Mussels

Purple cat's paw pearly mussels are nearly extinct. Dams on rivers where they lived in Ohio, Indiana, Illinois, Kentucky, Tennessee, and Alabama eliminated their sand and gravel river beds. Few of the mussels remain, and they no longer reproduce.

Mussels are cousins of oysters, scallops, and clams. Many live in the oceans, but a thousand species worldwide live in lakes and rivers, like the purple cat's paw. All have two-section shells connected by a hinge at the back, which the animal can open and close. The purple cat's paw is about the size of a man's palm, with wavy green rays on its shell. But a mussel is more than a shell.

MUSSELS

Inside the shell is the animal itself. It has a foot it uses as a digger for burrowing. The mussel also attaches itself to rocks with horny threads produced by a gland in its foot. The mussel has a proboscis, a bit like an elephant's trunk, that it can poke outside to feel about for food, which it then passes back inside to its mouth. Mussels eat decaying organic matter that otherwise would pollute streams and lakes.

Mussels have no heads, but they do have hearts that pump blood. They also have eyelike patches of tissue that sense light.

North American Indians once ate freshwater mussels. We now feed them to poultry and livestock. Factories turn their shells into buttons. Mussels in the Mississippi Valley have even produced pearls.

Dams, channel dredging, commercial mussel fishing, and pollution have endangered many freshwater mussels, like the Ouachita rock-pocketbook mussel of Oklahoma and Arkansas and the fanshell mussel of Kentucky, Tennessee, and Virginia. Purple cat's paw pearly mussels are almost gone. Biologists, who know of only one colony in Tennessee and one in Kentucky, say the species will probably soon become extinct.

34

Saving Wetlands

■ An egret is dancing.

Gray wings outstretched for balance, the big bird reels through shallow waters beside a small island in Florida's Tampa Bay. It only seems to dance; the egret actually is gorging on swarming killifish that swim in the waters of this unusual island, called Sunken Island.

Not long ago Sunken Island was barren. Now it is popping with life: sea grasses, mangrove thickets, raccoons, fish, crabs, birds. They are there because one biologist worried about wetlands.

Wetlands are places where the land is either soggy or covered with shallow water. A soggy wetland is a *bog*. Land covered with water, where mostly grasses grow, is a *marsh*. If trees grow from the water, it is a *swamp*. Such wetlands are rich habitats—or homes—for wildlife. Many endangered species depend on wetlands. But wetlands are disappearing fast.

In the 300 years since Europeans settled North America, the continent has lost up to 80 percent of its wetlands. Florida's Tampa Bay shows how it happened.

One hundred years ago, when Tampa's first settlers came, marshes and mangrove swamps rimmed the bay. Seagrass covered the bottom in vast underwater prairies, where clouds of shrimp and crabs grazed. The bay teemed with tarpon, redfish, snook, turtles, and manatees. Water birds filled the air. Captains sometimes stopped their ships because the water was so thick with fish. Since then, Tampa has spread, filling in the

MARSH

shore's wetlands with buildings and roads and polluting the water. Now, half the wetlands are gone.

A Tampa biology professor, Robin Lewis, worried about the wetlands. Then he learned of plans to dig channels for ships into the bay's bottom. A floating machine called a dredge does the digging. Lewis persuaded the dredgers to try an experiment. He knew humans could destroy marshes, but could they create one?

Years before, dredges had dug up sand from the bay's bottom and piled it at one spot in the water, creating Sunken Island. Now Robin Lewis arranged for the dredges to dump their sand at Sunken Island's tip.

Lewis calculated that currents would stretch out the pile of sand and curl it around to create a protected cove. He was right. The new cove was a perfect site for a marsh.

Bulldozers sloped the sides of the new cove. Then Lewis enlisted teenagers from a program for juvenile offenders, who do public services instead of going to jail. Lewis and his helpers planted spartina grass in the sand, where the seawater ebbed in and out with the tides. At first, the limp clumps of spartina looked glum, straggling across the beach. But within two years, thick grasses covered the sand, waving healthily in the breezes. Seeds of other plants floated in and took root. Now, 10 years later, the artificial cove is transformed. It is a wetland.

Recently, Robin Lewis took two visitors out in a boat to see his creation. Cabbage palms crowned the high ground. Lawnlike paspalum grass covered the upper beach, giving way to a thick marsh of belt-high spartina grass. Small fish jumped and glittered in the shallows, where herons waded. White and black mangrove trees now create a jungle along the edge of the upper beach, and red mangroves are colonizing the watery edges. Robin Lewis loves to see the mangroves growing because they cleanse the water of Tampa Bay. Their roots slow the motion of the water so that it drops the bits of polluting matter it carries. Then the roots absorb the pollutants.

The mangroves drop their twigs, seeds, roots, leaves, and fruits into the water. These droppings become coated with tiny mushroomlike plants, called fungi, and bacteria, which crabs and shrimp eat. Small fish eat the crabs and shrimp, and they are eaten, in turn, by larger fish, like tarpon and snook. Birds nest in the mangroves and eat the fish.

Over 1,000 pairs of birds now nest in Robin Lewis's artificial cove: glossy and white ibis, cattle egrets, snowy egrets, tri-colored herons, little blue herons, black-crowned and yellow-crowned night herons, and great egrets. None of these species is endangered as yet. If they have wetlands, none may ever be.

Since creating the cove on Sunken Island, Robin Lewis has planted many other marshes, mangrove swamps, and underwater seagrass meadows. At one spot along Tampa Bay, he transformed an abandoned garbage dump into a thriving marsh and mangrove swamp. One day there he counted 120 wood storks, large, white wading birds with naked black heads. Lewis was elated because wood storks are endangered.

It is the loss of their wetlands habitat that has endangered wood storks. By creating new marshes and swamps, "restoration ecologists" like Robin Lewis may help save them.

Other restoration ecologists are creating new prairies in the midwest, forests in Scotland, and tropical forests in Costa Rica. They would prefer to save natural habitats before they are lost. But humans have destroyed so many habitats that, for some species, creating a new habitat may be the best hope.

35

Ten Things You Can Do to Help Endangered Species

1. Clean Up Litter

When you visit a beach, pick up litter and throw it in a trash barrel. Here's why: Litter attracts rats and other animals that eat the eggs of shore birds.

Be especially careful of *plastic six-pack holders*. Plastic rings that hold cans and bottles of soda and other drinks are dangerous to wildlife, especially in lakes, streams, and the ocean. That is because the plastic six-pack holders are invisible in the water. Birds and animals become tangled in them and often die. People leave them on the beach. They also blow into the water from landfills. At home, cut apart your own plastic six-pack holders before you throw them out. When you find them littering shores, take them away and discard them properly.

Another wildlife killer is *balloons* that drift away and come down in water where species like whales and turtles think they are food. Do not release helium balloons—even balloons let go in the middle of the country can be blown for hundreds of miles, until they land in the ocean.

2. Avoid Styrofoam

Styrofoam is the lightweight plastic used in many throwaway cups and fast-food restaurant containers. Some styrofoam contains chemicals that damage the atmosphere. All styrofoam can be dangerous for such creatures as sea turtles. When it floats in the waves, styrofoam breaks into particles that animals think is food. If they eat it, they can die. Don't use styrofoam picnic plates and cups. At fast-food restaurants, see if they will give you paper plates instead of styrofoam packages.

3. *Support Wildlife Organizations*

You might persuade your class or club or friends to raise money to donate to one of the many organizations that look out for wildlife. You can raise money with bottle drives (if you live in a state with bottle deposits). You also can wash cars, mow lawns, rake leaves, or provide other services. Many organizations help wildlife; your school librarian may be able to help you find a complete list by looking in a directory of organizations and associations. Here are just a few groups that work to save endangered species:

- African Wildlife Foundation, 1717 Massachusetts Ave., NW, Washington, D.C. 20036
- Defenders of Wildlife, 1244 19th St., NW, Washington, D.C. 20036
- Environmental Defense Fund, 257 Park Ave. South, New York, NY 10010
- Friends of the Sea Otter, P.O.Box 221220, Carmel, CA 93922
- Greenpeace, 1436 U St., NW, Washington, D.C. 20009
- Humane Society of the U.S., 2100 L St., NW, Washington, D.C. 20037 (For kids, the Humane Society Youth Education Division offers the Kids In Nature's Defense Club.)
- International Crane Foundation, E-11376 Shady Lane Rd., Baraboo, WI 53913
- National Audubon Society, 950 Third Ave., New York, NY 10022

- National Parks and Conservation Association, 1015 31st St., NW, Washington, D.C. 20007
- National Wildlife Federation, 1400 16th St., NW, Washington, D.C. 20036
- The Nature Conservancy, 1800 North Kent St., Arlington, VA 22209
- Natural Resources Defense Council, 40 West 20th St., New York, NY 10011
- Rainforest Action Network, 466 Green St., San Francisco, CA 94133
- Sierra Club, 730 Polk St., San Francisco, CA 94109
- Whale Adoption Project, International Wildlife Coalition, P.O. Box 388, 634 North Falmouth Hwy., North Falmouth, MA 02556-0388
- Wilderness Society, 900 17th St., NW, Washington, D.C. 20077-7782
- Wildlife Conservation International, New York Zoological Society, Bronx, NY 10460
- Wildlife Preservation Trust International, 34th St. and Girard Ave., Philadelphia, PA 19104
- World Society for the Protection of Animals, P.O. Box 190, 29 Perkins St., Boston, MA 02130
- World Wildlife Fund, 1250 24th St., NW, Washington, D.C. 20037

4. Be a Zoo Volunteer

Call your local zoo to see if you can be a volunteer helper. See if your zoo is in charge of the recovery plan for an endangered species. You may be able to help.

5. Spread the Word

Post environmental tips on a bulletin board at school, church, synagogue, or club. See if you can set up an "environmental corner" at your local library for displays of environmental books, articles, and other information about the natural world and endangered species.

6. Watch Out for Marine Animals

If you live or vacation by the ocean, use boats, jet skis, and other motorized craft carefully in waters where sea mammals live. Whizzing boats that come too close can disturb whales, for instance. Manatees are frequently hurt or killed by boats that go too fast in waters where they live. Boat collisions can harm sea turtles, sea otters, seals, and other sea animals, too.

7. Be Careful What You Buy

Do not buy products made from wildlife, especially endangered species. For instance, elephants have been killed by the thousands just for their tusks. The ivory tusks are then made into products such as carved statues. Parrots and other birds sold as pets may have been captured in the wild; before you buy such a pet, especially endangered species, make sure it was bred in captivity and not taken from its wild home. Do not buy products made of tortoise shells, because it might be from endangered species. Certain types of fish nets accidentally kill large numbers of dolphins. When your family buys tuna at the supermarket, be sure the can says it is "dolphin safe." That means the tuna were caught using nets that do not harm dolphins.

8. Write Letters

You can write letters to your senators and representatives in Congress, asking them to support laws that protect the environment and wildlife. Your librarian can help you find their names and addresses. Write to an organization such as the Audubon Society, National Wildlife Federation, Wilderness Society, or Sierra Club to find out about environmental bills currently before Congress. Then, when you write to your representatives and senators you can mention those bills in particular.

9. Create a Bird Haven

Songbirds are declining. One reason is that as the human population expands, we take away habitat where the birds used to feed and nest. You can help by maintaining a bird feeder outside your house or apartment. If you have a lawn, you may want to plant shrubs and other plants that provide food and nesting places for birds. Write to the National Wildlife Federation (1412 16th St., NW, Washington, D.C. 20036-2266) for information on their Backyard Wildlife Habitat Program.

10. Ask a Wildlife Expert to Talk to Your Class

Find out if a nature center, science museum, conservation organization, or university near you can provide a speaker on endangered species.

Index